AF158393

The publishing house tredition has created the series **TREDITION CLASSICS**. It contains classical literature works from over two thousand years. Most of these titles have been out of print and off the bookstore shelves for decades.

The book series is intended to preserve the cultural legacy and to promote the timeless works of classical literature. As a reader of a **TREDITION CLASSICS** book, the reader supports the mission to save many of the amazing works of world literature from oblivion.

The symbol of **TREDITION CLASSICS** is Johannes Gutenberg (1400 – 1468), the inventor of movable type printing.

With the series, tredition intends to make thousands of international literature classics available in printed format again – worldwide.

All books are available at book retailers worldwide in paperback and in hardcover. For more information please visit: www.tredition.com

tredition was established in 2006 by Sandra Latusseck and Soenke Schulz. Based in Hamburg, Germany, tredition offers publishing solutions to authors and publishing houses, combined with worldwide distribution of printed and digital book content. tredition is uniquely positioned to enable authors and publishing houses to create books on their own terms and without conventional manufacturing risks.

For more information please visit: www.tredition.com

Reform Cookery Book (4th edition) Up-To-Date Health Cookery for the Twentieth Century.

Mrs. (Jean Oliver) Mill

Imprint

This book is part of the TREDITION CLASSICS series.

Author: Mrs. (Jean Oliver) Mill
Cover design: toepferschumann, Berlin (Germany)

Publisher: tredition GmbH, Hamburg (Germany)
ISBN: 978-3-8491-7179-7

www.tredition.com
www.tredition.de

Copyright:
The content of this book is sourced from the public domain.

The intention of the TREDITION CLASSICS series is to make world literature in the public domain available in printed format. Literary enthusiasts and organizations worldwide have scanned and digitally edited the original texts. tredition has subsequently formatted and redesigned the content into a modern reading layout. Therefore, we cannot guarantee the exact reproduction of the original format of a particular historic edition. Please also note that no modifications have been made to the spelling, therefore it may differ from the orthography used today.

"We could live without poets, we could live without books, But how in the world could we live without cooks."

PREFACE TO FOURTH EDITION.

Still the Food Reform movement goes on and expresses itself in many ways. New developments and enterprises on the part of those engaged in the manufacture and distribution of pure foods are in evidence in all directions. Not only have a number of new "Reform" restaurants and depots been opened, but vegetarian dishes are now provided at many ordinary restaurants, while the general grocer is usually willing to stock the more important health foods.

Then the interest in, and relish for a non-flesh dietary has, during the past year, got a tremendous impetus from the splendid catering at the Exhibitions, both of Edinburgh and London. The restaurant in Edinburgh, under the auspices of the Vegetarian Society, gave a magnificent object lesson in the possibility of a dietary excluding fish, flesh, and fowl. The sixpenny dinners, as also the plain and "high" teas, were truly a marvel of excellence, daintiness, and economy, and the queue of the patient "waiters," sometimes 40 yards long, amply testified to their popularity.

One is glad also to see that "Health Foods" manufacturers are, one after another, putting into practice the principle that sound health-giving conditions are a prime essential in the production of what is pure and wholesome, and in removing from the grimy, congested city areas to the clean, fresh, vitalising atmosphere of the country, not only the consumers of these goods, but those who labour to produce them, derive real benefit.

The example of Messrs Mapleton in exchanging Manchester for Wardle, has been closely followed up by the International Health Association, who have removed from Birmingham to Watford, Herts.

J. O. M.

NEWPORT-ON-TAY, *April 1909.*

"Economy is not Having, but wisely spending." *Ruskin*.

"I for my part can affirm that those whom I have known to submit to this (the vegetarian) regimen have found its results to be restored or improved health, marked addition of strength, and the acquisition by the mind of a clearness, brightness, well-being, such as might follow the release from some secular, loathsome detestable dungeon.... All our justice, morality, and all our thoughts and feelings, derive from three or four primordial necessities, whereof the principal one is food. The least modification of one of these necessities would entail a marked change in our moral existence. Were the belief one day to become general that man could dispense with animal food, there would ensue not only a great economic revolution — for a bullock, to produce one pound of meat, consumes more than a hundred of provender — but a moral improvement as well." — *Maurice Maeterlinck*.

"Can anything be so elegant as to have few wants, and to serve them one's self, so to have somewhat left to give, instead of being always prompt to grab." — *Emerson*.

Foreword.

"Diet cures mair than physic." — *Scotch Proverb.*
"The first wealth is health." — *Emerson.*

"Of making books there is no end," and as this is no less true of cookery books than of those devoted to each and every other subject of human interest, one rather hesitates to add anything to the sum of domestic literature. But while every department of the culinary art has been elaborated *ad nauseam*, there is still considerable ignorance regarding some of the most elementary principles which underlie the food question, the relative values of food-stuffs, and the best methods of adapting these to the many and varied needs of the human frame. This is peculiarly evident in regard to a non-flesh diet. Of course one must not forget that there are not a few, even in this age, to whom the bare idea of contriving the daily dinner, without the aid of the time-honoured flesh-pots, would seem scarcely less impious than absurd, as if it threatened the very foundations of law and order. Still there is a large and ever increasing number whose watch word is progress and reform, who would be only too glad to be independent of the *abattoir* (I will not offend gentle ears with the coarse word slaughter-house), if they only knew how. In summertime, at least, when animal food petrifies so rapidly, many worried housekeepers, who have no prejudice against flesh-foods in general, would gladly welcome some acceptable substitute. The problem is how to achieve this, and it is with the view of helping to that solution that this book is written.

Now, as I said, while there is no lack of the stereotyped order of domestic literature, there seems to be a wide field over which to spread the knowledge of "Reform" dietary, and how to adapt it to the needs of different people, and varying conditions. And while protesting against all undue elaboration — for all true reform should simplify life rather than complicate it—we should do well to acquire the knowledge of how to prepare a repast to satisfy, if need be, the most exacting and fastidious.

Another need which I, as a Scotswoman, feel remains to be met, is a work to suit the tastes and ideals of Scottish people. Cosmopolitan as we now are, there are many to whom English ways are unfamiliar. Even the terms used are not always intelligible, as is found by a Scotswoman on going to live in England, and *vice-versa*. We could hardly expect that every London stoneware merchant would be able to suit the Scotch lass, who came in asking for a "muckle broon pig tae haud butter;" but even when English words are used, they may convey quite different ideas to Scottish and English minds. Indeed, several housewives have complained to me that all the vegetarian cookery books, so far as they can learn, are intended solely for English readers, so that we would hope to overcome this difficulty and yet suit English readers as well.

Before starting to the cookery book proper, I would point out some of the commonest errors into which would-be disciples of food reform so often fall, and which not unfrequently leads to their abandoning it altogether as a failure. Nothing is more common than to hear people say most emphatically that vegetarian diet is no good, for they "have tried it." We usually find upon enquiry, however, that the "fair trial" which they claim to have given, consisted of a haphazard and ill-advised course of meals, for a month, a week, or a few days intermittently, when a meat dinner was from some reason or other not available. One young lady whom I know, feels entitled to throw ridicule on the whole thing from the vantage-ground of one day's experience—nay, part of a day. It being very hot, she could not tackle roast beef at the early dinner, and resolved with grim heroism to be "vegetarian" for once. To avoid any very serious risks, however, she fortified herself as strongly as possible with the other unconsidered trifles—soup, sweets, curds and cream, strawberries, &c., but despite all her precautions, by tea-time the aching void became so alarming that the banished joint was recalled from exile, and being "so famished" she ate more than she would have done at dinner. Next day she was not feeling well, and now she and her friends are as unanimous in ascribing her indisposition to vegetarianism, as in declaring war to the knife—or *with* the knife against it evermore.

Now, there are certainly not many who would be so stupid or unreasonable as to denounce any course of action on the score of one spasmodic attempt, but there are not a few who are honestly desirous to follow out what they feel to be a better mode of living, who take it up in such a hasty, ill-advised way as to ensure failure. It is not enough merely to drop meat, and to conclude that as there is plenty food of some or any sort, all will be right, unless it has first been ascertained that it will contain the essential elements for a nourishing, well-balanced meal. It is not the quantity, however, which is so likely to be wrong as the proportions and combination of foods, for we may serve up abundance of good food, well cooked and perfectly appointed in every way, and yet fail to provide a satisfactory meal. I would seek to emphasise this fact, because it is so difficult to realise that we may consume a large amount of food, good in itself, and yet fail to benefit by it. If we suffer, we blame any departure from time-honoured orthodoxy, when, perhaps we ought to blame our wrong conception or working out of certain principles. It is never wise, therefore, to adopt the reform dietary too hastily, unless one is quite sure of having mastered the subject, at least in a broad general way; for if the health of the household suffers simultaneously with the change, we cannot hope but that this will be held responsible. Other people may have "all the ills that flesh is heir to" as often as they please. A vegetarian dare hardly sneeze without having every one down upon him with 'I told you so.' 'That's what comes of no meat.'

A frequent mistake, then, is that of making a wrong selection of foods, or combining them unsuitably, or in faulty proportions. For example, rice, barley, pulses, &c., may be, and are, all excellent foods, but they are not always severally suitable under every possible condition. Rice is one of the best foods the earth produces, and probably more than half of the hardest work of the world is done on little else, but those who have been used to strong soups, roast beef, and plum pudding will take badly with a sudden change to rice soups, rice savoury, and rice pudding. For one thing, so convinced are we of the poorness of such food, that we should try to take far too much, and so have excess of starch. Pulse foods, again,—peas, beans, lentils—are exceedingly nutritious—far more so than they get credit for, and in their use it is most usual to heavily overload the system with excess of nitrogenous matter. One lady told me she understood one had to take enormous quantities of haricot beans, and she was quite beat to take *four* platefuls! 'I can never bear the sight of them since,' she added pathetically. Another—a gentleman—told me vegetarianism was 'no good for him, at any rate, for one week he swallowed "pailfuls of swill," and never felt satisfied!' While yet a third—no, it was his anxious wife on his behalf—complained that 'he could not take enough of "that food" to keep up his strength.' He had three platefuls of the thickest soup that could be contrived, something yclept "savoury"—though I cannot of course vouch for the accuracy of that definition—a substantial pudding, and fruit. He 'tried' to take two tumblers of milk, but despite his best endeavours could manage to compass only *one*! I sympathised heartily with the good lady's anxiety, and urged that they go back to their "morsel of meat" without delay, and dispense with the soup, the "savoury," the milk, and either the fruit or the pudding. In reply to her astonished look, I gravely assured her that it was evident vegetarianism would not do for them, and her look of relief made it clear that she never suspected the mental reservation, that the tiny bit of meat was invaluable if only to keep people from taking so much by way of compensation.

Another mistake to be guarded against, is that of reverting too suddenly to rather savourless insipid food. It is certainly true that as one perseveres in a non-flesh diet for a length of time, the relish for spices and condiments diminishes, and one begins to discern new, subtle, delicate flavours which are quite inappreciable when accustomed to highly seasoned foods. As one gives up these artificial accessories, which really serve to blunt the palate, rarer and more delicious flavours in the sweet natural taste come into evidence. But this takes time. There is a story told of some Londoners who went to visit at a country farm, where, among other good things, they were regaled with new-laid eggs. When the hostess pressed to know how they were enjoying the rural delicacies, they, wishing to be polite yet candid, said everything was very nice, but that the eggs had not "the flavour of London ones!"

It were thus hopeless to expect those who like even eggs with a "tang" to them, to take enthusiastically to a dish of tasteless hominy, or macaroni, but happily there is no need to serve one's apprenticeship in such heroic fashion. There is at command a practically unlimited variety of vegetarian dishes, savoury enough to tempt the most fastidious, and in which the absence of "carcase" may, if need be, defy detection. Not a very lofty aspiration certainly, but it may serve as a stepping-stone.

When the goodman, therefore, comes in expecting the usual spicy sausage, kidney stew, or roast pig, do not set before him a dish of mushy barley or sodden beans as an introduction to your new 'reform bill' of fare, or there may be remarks, no more lacking in flavour than London eggs. Talking of sausage, reminds me that one of the favourite arguments against vegetarian foods is that people like to know what they are eating. What profound faith these must have in that, to us cynical folks, 'bag of mystery,' the sausage! But then, perhaps, they do know that they are eating— —!

Now, I fear most of the foregoing advice on how to "Reform" sounds rather like Punch's advice to those about to marry, so after so many "don'ts" we must find out how to *do*. And to that end I would seek rather to set forth general broad guiding principles instead of mere bald recipes. Of course a large number of the items—puddings, sweets, &c., and not a few soups, are the same as in ordinary fare, so that I will give most attention to savouries, entrees, and the like, which constitute the real difficulty.

As people get into more wholesome ways of living, the tendency is to have fewer courses and varieties at a meal, but just at first it may be as well to start on the basis of a three-course dinner. One or other of the dishes may be dispensed with now and then, and thus by degrees one might attain to that ideal of dainty simplicity from which this age of luxury and fuss and elaboration is so far removed.

"Now good digestion wait on appetite,
And health on both." — *Shakespeare.*

SOUPS.

The following directions will be found generally applicable, so that there will be no need to repeat the several details each time. Seasonings are not specified, as these are a matter of individual taste and circumstance. Some from considerations of health or otherwise are forbidden the use of salt. In such cases a little sugar will help to bring out the flavour of the vegetables, but unless all the members of the household are alike, it had best not be added before bringing to table. Where soup is to be strained, whole pepper, mace, &c., is much preferable to ground, both as being free from adulteration, and giving all the flavour without the grit. The water in which cauliflower, green peas, &c., have been boiled, should be added to the stock-pot, but as we are now recognising that all vegetables should be cooked as conservatively as possible — that is, by steaming, or in just as much water as they will absorb, so as not to waste the valuable salts and juices, there will not be much of such liquid in a "Reform" menage. A stock must therefore be made from fresh materials, but as those are comparatively inexpensive, we need not grudge having them of the freshest and best. Readers of Thackeray will remember the little dinner at Timmins, when the hired *chef* shed such consternation in the bosom of little Mrs Timmins by his outrageous demands for 'a leg of beef, a leg of veal, and a ham', on behalf of the stock-pot. But the 'Reform' housekeeper need be under no apprehension on that score, for she can have the choicest and most wholesome materials fresh from the garden to her *pot-au-feu*, at a trifling cost. Of course it is quite possible to be as extravagant with vegetarian foods as with the other, as when we demand forced unnatural products out of their season, when their unwholesomeness is matched only by their cost. No one who knows what sound, good food really is, will dream of using manure-fed tomatoes, mushrooms at 3s. per lb.; or stringy tough asparagus, at 5s. or 10s. a bunch, when seasonable products are to be had for a few pence.

The exact quantities are not always specified either, in the following recipes, as that too has to be determined by individual requirement, but as a general rule they will serve four to six persons. The amount of vegetables, &c., given, will be in proportion to 3 pints, i.e. 12 gills liquid. Serve all soups with croutons of toast or fried bread.

White Stock.

The best stock for white soups is made from small haricots. Take 1 lb. of these, pick and wash well, throwing away any that are defective, and if there is time soak ten or twelve hours in cold water; put on in clean saucepan — preferably earthenware or enamelled — along with the water in which soaked (if not soaked scald with boiling water, and put on with fresh boiling water), some of the coarser stalks of celery, one or two chopped Spanish onions, blade of mace, and a few white pepper-corns. If celery is out of season, a little celery seed does very well. Bring to boil, skim, and cook gently for at least two hours. Strain, and use as required.

Clear Stock.

For clear stock take all the ingredients mentioned above, also some carrot and turnip in good-sized pieces, some parsley, and mixed herbs as preferred, and about 1/2 lb. of hard peas, which should be soaked along with the haricots. Simmer very gently two to three hours. Great care must be taken in straining not to pulp through any of the vegetables or the stock will be muddy, or as we Scotch folks would say "drumlie." If not perfectly clear after straining, return to saucepan with some egg-shells or white of egg, bring to boil and strain again through jelly-bag. A cupful of tomatoes or a few German lentils are a great improvement to the flavour of this stock, but will of course colour it more or less.

Brown Stock.

Take 1/2 lb. brown beans, 1/2 lb. German lentils, 1/2 lb. onions, 1 large carrot, celery, &c. Pick over the beans and lentils, and scald for a minute or two in boiling water. This ensures their being perfectly clean, and free from any possible mustiness. Strain and put on with fresh boiling water some black and Jamaica pepper, blade mace, &c., and boil gently for an hour or longer. Shred the onion, carrot, and celery finely and fry a nice brown in a very little butter taking great care not to burn, and add to the soup. Allow all to boil for one hour longer, and strain. A few tomatoes sliced and fried along with, or instead of the carrot, or a cupful of tinned tomatoes would be a great improvement. This as it stands is a very fine

Clear Brown Soup,

but if a thicker, more substantial soup is wanted, rub through as much of the pulp as will give the required consistency. Return to saucepan, and add a little soaked tapioca, ground rice, cornflour, &c., as a *liaison*. Boil till that is clear, stirring well. Serve with croutons of toast or fried bread. This soup may be varied in many ways, as by adding some finely minced green onions, leeks, or chives either before or after straining and some parsley a few minutes before serving.

White Windsor Soup.

Take 4 breakfast cupfuls white stock or water, add 6 tablespoonfuls mashed potato and 1 oz fine sago. Stir till clear and add 1 breakfast cup milk and some minced parsley. Let come just to boiling point but no more. If water is used instead of stock some finely shred onion should be cooked without browning in a little butter and added to the soup when boiling. Rub through a sieve into hot tureen.

White Soubise Soup.

Melt in lined saucepan 2 oz. butter, and into that shred 1/2 lb. onions. Allow to sweat with lid on very gently so as not to brown for about half an hour. Add 1-1/2 pints white stock and about 6 ozs. scraps of bread any hard pieces will do, but no brown crust. Simmer very gently for about an hour, run through a sieve and return to saucepan with 1 pint milk. Bring slowly to boiling point and serve. To make

Brown Soubise Soup

toast the bread, brown the onions, and use brown stock.

Almond Milk Soup.

Wash well 1/4 lb. rice and put on to simmer slowly with 1-1/2 pints milk and water, a Spanish onion and 2 sticks of white celery. Blanch, chop up and pound well, or pass through a nut-mill 1/4 lb. almonds, and add to them by degrees another 1/2 pint milk. Put in saucepan along with some more milk and water to warm through, but do not boil. Remove the onion and celery from the rice (or if liked they may be cut small and left in), and strain the almonds through to that. See that it is quite hot before serving.

NOTE.—For this and other soups which are wanted specially light and nourishing, Mapleton's Almond Meal will be found exceedingly useful. It is ready for use, so that there is no trouble blanching, pounding, &c.

Brazil Soup.

Put 1 lb. Brazil nuts in moderate oven for about 10 minutes, remove shells and brown skin—the latter will rub off easily if heated—and grate through a nut-mill. Simmer gently in white stock or water with celery, onions, &c., for 5 or 6 hours. Add some boiling milk, pass through a sieve and serve. A little chopped parsley may be added if liked.

Chestnut Soup.

Chop small a good-sized Spanish onion and sweat in 1 oz. butter for twenty minutes. Add 2 to 3 pints stock and 1 lb. chestnuts previously lightly roasted and peeled. Simmer gently for one hour or more, pass through a sieve and return to saucepan. Bring to boil, remove all scum, add a cupful boiling milk or half that quantity of cream, and serve without allowing to boil again.

Plain White Soup.

Into enamelled saucepan put 2 ozs. butter, and as it melts stir in 2 ozs. flour. Add very gradually a breakfast cup milk, and stir over a slow heat till quite smooth. Add 3 or 4 breakfast cupfuls white stock, bring slowly to boil and serve.

Velvet Soup.

Prepare exactly as for Plain White Soup, but just before serving beat up the yokes of 2 or 3 eggs. Add to them a very little cold milk or cream, and then a little of the soup. Pass through strainer into hot tureen, strain through the rest of the soup, and mix thoroughly.

Parsnip Soup.

Take 1/2 lb. cooked parsnips or boil same quantity in salted water till tender, pass through a sieve and add to a quantity of Plain White Soup or Stock. Bring to boil, and if sweet taste is objected to add strained juice of half a lemon.

Turnip Soup.

is made in exactly the same way as Parsnip Soup, substituting young white turnips or "Golden Balls" for the parsnips, and many people will prefer the flavour. A little finely chopped spring onion or chives and parsley would be an improvement to both soups. These—except the parsley—should be boiled separately and added just before serving.

Palestine Soup.

A very fine soup is made thus:—Pare and boil 2 lbs. Jerusalem Artichokes in milk and water with a little salt till quite soft, then pass through a sieve or potato masher, and add to quantity required of Velvet Soup.

Westmoreland Soup.

Put in soup pot some very plain stock, or water will do quite well. Add 1 lb. lentils, 1/2 lb. onions, small carrot, piece of turnip, and a stick or two of celery, all chopped small, also a teacupful tomatoes. Boil slowly for two hours, pass through a sieve and return to soup pot. Melt a dessert-spoonful butter and stir slowly into it twice as much flour, add gradually a gill of milk. When quite smooth add to soup and stir till it boils.

This is a very good soup and might be preferred by some without straining the vegetables. The lentils might be boiled separately and put through a sieve before adding.

The foregoing are all varieties of White Soup and these could be extended indefinitely; but as such variations will suggest themselves to everyone, it is not necessary to take up space here. I might just mention that a most delicious

Cauliflower Soup

can be made by adding a nice young cauliflower, all green removed, cut in tiny sprigs, and boiled separately to the quantity required of Plain White Soup. The water in which boiled should be added also.

White Haricot Soup

is made by substituting haricot or butter beans for the cauliflower. These should be slowly cooked till tender and passed through a sieve or masher.

Celery Soup.

For this use a large well-blanched head of celery. Either chop small when cooked, or pass through sieve before adding to White Soup.

Asparagus Soup.

Take a bunch tender asparagus. Set aside the tops. Blanch stalks in salted boiling water for a minute or two, then drain and simmer till tender in a little milk and water. Pulp through sieve and add to White Soup when boiling. Cook the tops separately in salted boiling water. Drain and add to soup in tureen. Tinned asparagus makes very good soup. It requires little or no cooking, only to be made quite hot. Pulp stalks and put in tops whole.

Clear Soups.

It is unnecessary to give every recipe in detail for these also, if a rich clear stock has been prepared according the directions, page 11. These of course may be varied according to taste or convenience, and all the ingredients specified are by no means indispensable. Some may be left out and others added as they are at hand or in season. When celery is not to be had celery seed or celery salt gives a good flavour. A hasty stock may be contrived at anytime with chopped onions, shred carrot, and some lentils—green or yellow or both. The vegetables should be lightly fried in a little butter, the lentils scalded or washed well, and all boiled together for an hour or even less with the required quantity of water. Strain without any pressure. Then a still more hasty stock can be had with any of the excellent "Extracts" which are on the market. Their flavour will be appreciated by all, and the fact that they are manufactured from pure, wholesome cereals—barley, chiefly, I believe—should go a long way to commend them to those who have no favour for the uric acid products of "Animal" Extracts.

Well, then, if a good, clear stock is prepared, all that is necessary to convert it into

Clear Soup a la Royale

is to prepare a savoury custard with two yolks and either a cup of stock, diluted "Extract," or milk. Steam in shallow, buttered tin, cut in small squares, diamonds, &c., and put in tureen along with the boiling stock.

Julienne Soup.

Cut different vegetables—carrot, turnip, celery, &c., in thin strips about 1 inch long, boil in salted water, and add to boiling clear stock.

Spring Vegetable Soup.

Have an assortment of different young vegetables comprising as many distinct and bright colours as possible—green peas, French beans trimmed and cut diamond-wise, cauliflower in tiny sprigs, carrots, turnips, cooked beetroot stamped in fancy shapes or cut in small dice, and leeks, chives, or spring onions shred finely. Cook the vegetables separately, drain, and add while hot to boiling clear stock in tureen.

Thick Soups.

Most of the thick soups are so well-known that they need not be repeated here. Suffice it to say that they will gain both in purity and flavour by substituting vegetarian stock for that usually made by boiling meat, ham bones, and the like. Great care should be taken with such soups as lentil, split-pea, potato soup, &c., to avoid a coarse "mushy" consistency. This can be done by rubbing the peas, &c., through a sieve when cooked, and adding such vegetables as carrot, turnip, onions, &c., finely chopped, to the strained soup. Perhaps, however, I ought to give at least one typical recipe for

"Reform" Pea Soup,

and if nicely made it will be quite possible to allure some unsuspecting victims who have always declared they never could or would touch pea soup, into asking for another helping of "that delicious—ahem—what-do-you- call-it-soup."

Have ready a good-sized-soup pot with amount of water required boiling fast, and into this throw 1/2 lb. split-peas for every 2 pints water. The "Giant" variety is best as they are BO easily examined and cleaned. Rub in a coarse cloth to remove any possible dust or impurity. This is much better than washing or scalding, as the peas "go down" so much more quickly when put dry into the fast boiling water. Such a method will seem rather revolutionary to those who have been accustomed to soak peas over night, but a single trial is all that is needed to convince the most sceptical. Add 1/2 lb. onions, cut up-these may first be sweated for 10 minutes with a little butter in covered pan. Simmer gently but steadily 1 1/2 to 2 hours. Rub through a sieve and return to saucepan. When boiling add some turnip in tiny dice and some carrot in slices as thin as sixpence, also finely chopped spring onion, leeks or chives, according to season, and a little finely minced parsley five minutes before serving. Stock may of course be used for this soup, but is not at all necessary. With stock or even a little extract, a very good lentil or pea soup may be made at a few minutes' notice by thickening with

"Digestive" Pea Flour

or lentil flour, as the case may be. Such soups can be taken by those of weak digestion. No vegetables should be added in that case, or if so they should be strained out.

Mulligatawny Soup.

Chop up 2 apples and 1 Spanish onion and stir over the fire with 2 ozs. butter till quite brown, but not burnt. Add 1 oz. flour (and if wanted somewhat thickened, one or two spoonfuls "Digestive" lentil or pea flour), 1 teaspoonful curry powder, and a cupful of milk, previously mixed together. Stir till smooth and boil up, then add some good stock—brown would be best—and simmer for half an hour longer, removing the scum as it rises. Serve with boiled rice, handed round on a separate dish.

Hotch-Potch.

This soup is to be had in perfection in the summer months when young, tender vegetables are to be had in great variety and abundance. The more different kinds there are the better, but care must be taken to give each just the proper amount of cooking and no more, or the result will be that by the time certain things are done, others will be mushy and insipid. Bring to boil the necessary quantity of clear stock—water will do. Have ready a cupful each of carrots and turnips in tiny dice—the smaller ends of the carrots being in thin slices—a cauliflower in very small sprigs, one or two crisp, tender lettuces finely shred, cupful green peas, some French beans trimmed and cut small, a dozen or so of spring onions, 2 tablespoonfuls each of lentils and rice, and any other seasonable vegetable that is to be had. Add each in their turn to the boiling stock, the time required being determined by age and condition. If very young and fresh, the carrots will require only 30 to 40 minutes, the turnips and spring onions rather less, and the cauliflower less still. French beans require about 20 minutes, peas and lettuce 15 minutes, while the rice and lentils should have about half an hour. Much must be left to the discretion of the cook, but one point I would emphasise is, don't over-boil the vegetables. There seems to be an idea that a safe rule for vegetables is the more you cook them the better, but the fact is they lose in flavour and wholesomeness every five minutes after they are done. This is why "second day's" soup so often disagrees when the first has been all right. A few slices of tomato may be added. They should be fried in a little butter, cut small, and added shortly before serving, also some chopped parsley.

Winter Hotch-Potch.

This also may be very good. All the vegetables will require much longer cooking. Some will not be available, but in their place will be celery, parsnips, Brussels sprouts, leeks, &c. Dried green peas, soaked for 12 hours, can be used, or a good canned variety, and I may say that many delicious vegetables are now to be had in tins, or, better still, in glass jars.

Scotch Broth.

For this wash well a cupful good fresh *pot* barley, bring to boil in plenty of water, pour that off and put on with clean cold water. Simmer for 2 hours and then add a selection of vegetables given for Hotch-Potch.

Mock Cock-a-Leekie

or Leek Soup (*maigre*) is an excellent winter soup. Take a dozen or more crisp fat leeks—flabby, tough ones are no use—trim away all coarse pieces, chop up the tender green quite small and simmer in covered pan with a little butter. Add to quantity required of either white stock or plain white soup, which should be boiling. Shred down the white of the leeks, fry in a little more butter, and add twenty minutes later. Cook till quite tender. If stock is used, some well-washed rice should be added about 30 minutes before serving. If white soup is prepared, it is best to cook the leeks thoroughly before adding, then merely bring to boil and serve.

Green Pea Soup.

This is a delicious summer soup. Have a clear stock made with fresh green vegetables, such as lettuce, green onions, spinach, bunch parsley, sprig mint, &c., the shells wiped clean and about half of the peas—about 2 lbs. will be needed—reserving the finest. Rub through a sieve, return to saucepan and bring to boil. Add remainder of peas, boil 15 minutes, and pour into tureen over an ounce or so of butter. Some may prefer cream in place of butter, in which case add just before serving, and do not allow to boil up.

Mock Hare Soup.

Prepare a rich well-flavoured brown stock, rubbing through the greater part of the German lentils, &c., to make it of a thick creamy consistency. The flavour will be best if such vegetables as carrot and onion are sliced and fried brown before boiling. Toast two tablespoonfuls oatmeal and one of flour to a light brown, mix with it a teaspoonful ground Jamaica pepper and smooth with a little cold water. Add to the boiling soup and stir till it boils up again. Mushroom ketchup, a few fried mushrooms, some piquant sauce, "Extract," &c., &c., may be added or not at discretion.

German Lentil Soup.

Scald 1/2 lb. German lentils for a minute in boiling water, drain and put on with quantity of boiling water required. Fry some onions, celery, and tomatoes — if to be had — in a little butter till brown, and add. Simmer about 2 hours, and rub through a sieve. Add a little ground rice, cornflour, &c., to keep the pulp from settling to the bottom. A little milk or cream or ketchup may be added if liked.

Butter Peas Soup.

Cook butter peas as for stew, [Footnote: See page 35. [Butter Peas or "Midget" Butter Bean, below]] pulp through a sieve and add to quantity of liquid required, which may be white stock or milk and water, and should be boiling. Add a small white cauliflower, cut in tiny sprigs (or any tender fresh vegetables cut small and parboiled separately). Simmer till cauliflower is just cooked, add some chopped parsley, and serve.

Mock Turtle Soup.

Prepare a quantity of strong, clear, highly-flavoured stock of a greenish-brown colour. The colour can be obtained by boiling some winter greens or spinach along with the other things. A few chopped gherkins, capers, or chillies will give the required piquancy. Have 4 ozs. tapioca soaked overnight, add to the boiling stock and cook gently till perfectly clear. Some small quenelles may be poached separately and put in tureen.

Tomato Soup.

When this soup is well made it is a general favourite, but it must be well made, for it is impossible to appreciate the greasy, yellow, dish-water-looking liquid which is sometimes served in that name.

Put in a saucepan 2 ozs. butter, and into that shred finely 1/2 or 1 lb. onions. Add half or more of a tin of tomatoes or about 1 lb. fresh ones sliced, and a cup of water or stock. Simmer very gently for an hour and rub through a wire sieve, pressing with the back of a wooden spoon to get all the pulp through. *Everything* should go through except the skin and seeds. Return to clean saucepan with stock or water, and two tablespoonfuls of tapioca, previously soaked for at least an hour. Stir till it boils and is quite clear. This soup may be varied in many ways, as by substituting for the tapioca, crushed vermicelli, ground rice, cornflour, &c. Some chopped spring onions, chives or leeks, added after straining are a great improvement, also chopped parsley, while many people like the addition of milk or cream.

SAVOURIES.

"We live not upon what we eat, but upon what we digest."

We come now to consider the middle courses of dinner in which lies the crux of the difficulty to the aspirant who wishes to contrive such without recourse to the flesh-pots. This is where, too, we must find the answer to those half-curious wholly sceptical folks who ask us, "Whatever *do* you have for dinner?" Most of them will grant that we *may* get a few decent soups, though no doubt they retain a sneaking conviction that at best these are "unco wersh," and puddings or sweets are almost exclusively vegetarian. But how to compensate for that little bit of chicken, ox, or pig—no one now-a-days owns to taking much meat!—is beyond the utmost efforts of their imagination. Of course we can't have everything. When a "reformed" friend of mine was asserting that we could have no end of delicacies, one lady triumphantly remarked "Anyhow, you can't have a leg of mutton." That is true, but then we must remember that it's not polite to speak of "legs," especially with young ladies learning cooking. Liver or kidneys are not particularly nice things to speak about either, and I am sure if we reflected on what their place is in the economy of the body, we should think them still less nice to eat.

But joking apart, there is a growing tendency to get as far away as we can from their origin in the serving of meat dishes. The old-time huge joints, trussed hares, whole sucking pigs, &c., are fast vanishing from our tables, and the smart *chef* exerts himself to produce as many recherche and mysterious little made dishes as possible. Not a few of these are quite innocent of meat, indeed, that is the complaint urged against them by those who believe that in flesh only can we have proper sustenance. But little research is needed, however, to show that apart from flesh foods there are immense and only partially developed resources in the shape of cereals, pulses, nuts, &c.,

and, it is to these that we must look for our staple solid foods. In a small work like this it is impossible to do much more than indicate the lines upon which to go, but I shall try to give as many typical dishes as I can, and to suggest, rather than detail, variations and adaptations.

We must first study very briefly the various food elements, and learn the most wholesome and suitable combination of these. In an ordinary three-course dinner we must arrange to have a savoury that will fitly follow the soup and precede the sweets. Thus, if we have a light, clear, or white soup, we shall want a fairly substantial savoury, and if the soup has been rather satisfying it must be followed by a lighter course.

The lightest savouries are prepared mostly from starch foods, as rice, macaroni, &c., while for the richer and more substantial we have recourse to peas, beans, lentils, and nuts.

The first set of savouries given are of the lighter description, and are well suited to take the place of the fish course at dinner.

LIGHT SAVOURIES.

Fillets of Mock Sole.

Bring to boil 1/2 pint milk and stir in 2 ozs. ground rice or 3 ozs. flaked rice. Add 1 oz. butter, teaspoonful grated onion, and a pinch of mace. Add also three large tablespoonfuls of potato which has been put through a masher or sieve, mix, and let all cook for 10 to 20 minutes. As the mixture should be fairly stiff this can best be done in a steamer or double boiler. When removed from the fire add 1 egg and 1 yolk well beaten. Mix thoroughly and turn out on flat dish not quite 1/2 inch thick, and allow to get quite cold. Divide into fillet-shaped pieces, brush over with white of egg beaten up, toss in fine bread crumbs and fry in deep smoking-hot fat. Drain, and serve very hot, garnished with thin half or quarter slices of lemon, and hand round Dutch sauce in tureen.

Fillets of Artichoke.

Boil some Jerusalem Artichokes till tender, but not too soft, cut in neat slices, and egg, crumb, and fry as above.

Chinese Artichokes.

Salsify, Scorzonera, &c., may be done in same way. Serve with Dutch or tomato sauce. A variety is made by simply boiling or steaming in milk and water. Drain, and serve with parsley or other sauce poured over.

Celery Fritters.

Get a good-sized head of well-blanched celery, trim and cut in small pieces, put in salted boiling water for a few minutes, then drain. Into a stewpan, or much better a steamer or double boiler, put 1/2 oz. butter, and into that shred a very small Spanish onion or a few heads of spring onion or shallots. Add the drained celery, one or two spoonfuls milk, salt, white pepper, and pinch mace. Allow to cook till quite tender then pour over a slice of bread free from crust and crumbled down. If the bread is not moist enough add a little hot milk. Allow to stand for a time, then drain away any superfluous moisture. The difficulty is to get this dry enough, and that is why a double saucepan is much better than an open pan, in which it is scarcely possible to cook dry enough without burning. Make a sauce with 1/2 oz. butter, 1/2 oz. flour, and 1/2 gill milk, and when it thickens add the panada, celery, &c. Stir over gentle heat till the mixture is quite smooth and leaves the sides of the pan. Remove from the fire and mix in one or two beaten eggs. Turn out to cool, shape into fritters, and fry as mock sole.

Cauliflower Fritters

are made same as above, with cauliflower in place of celery.

Note.—The eggs in this and mock sole may be left out, though they are an improvement and help to bind the mixture together. Variety can be obtained by varying the seasonings, adding a little lemon juice or Tarragon vinegar, &c., either to the mixture or to the sauce.

Macaroni Omelet.

Boil 2 ozs. short cut macaroni in salted boiling water, and drain. Put 3 dessertspoonfuls flour in a basin, smooth with a little cold milk, and pour a breakfast-cupful boiling milk over it, stirring vigorously all the time. Add one or two spoonfuls of cream — or a little fresh dairy butter or nut butter beat to a cream — 2 beaten eggs, teaspoonful minced parsley, same of grated onion, the macaroni, a large cup bread crumbs, seasoning of pepper, salt, &c. Mix very well. Put in buttered pie-dish and bake 30 to 40 minutes in brisk oven. Turn out and serve with brown or tomato sauce. Some grated cheese may be added if liked.

Macaroni Cutlets.

Boil 3 or 4 ozs. macaroni in salted water for 15 minutes. Drain, and stew or steam till very tender along with some shred onion and tomatoes previously fried together, without browning, in 1 oz. butter. If too dry add a very little milk. When quite tender mix in enough bread crumbs to make a rather stiff consistency, also 1 or 2 ozs. grated cheese. Mix well over the fire. Add a beaten egg, pinch mace, and any other seasoning. Mix well again, turn out to cool, form into pear-shaped cutlets, egg, crumb, and fry in usual way.

Macaroni Egg Cutlets

are made by adding 2 finely chopped hard boiled eggs to the above mixture. Add when macaroni is cooked, along with crumbs, raw egg, seasoning, &c.

Celery Egg Cutlets

are made by adding the hard-boiled eggs to the mixture for celery fritters. Both of these are specially delicious, and this forms an excellent way of using up cold cooked stuff—savoury rice, vermicelli, &c.—so that one can have a dainty savoury with very little trouble. This is of no little importance in an age when so many demands are made upon the time and energy of the average housewife, and one would do well to study while preparing any dish requiring a good deal of care and labour, to have sufficient over to make a fricassee of some sort for another time.

Rice and Lentil Mould

comes in very handy in this way. Put 1 oz. butter in saucepan and shred into it very finely a large Spanish onion or an equal quantity white of small onions or leeks. Cover, and allow to sweat over gentle heat for 10 minutes. Some finely shred white celery along with the onions is a welcome addition, but is not indispensable. Pick and wash well 1/4 lb. yellow lentils and bring to boil in water to cover. Do the same with 3 ozs. rice. The lentils and rice may be boiled together, but are nicer done separately. Add to onion, &c., in saucepan, along with seasoning to taste of curry powder, &c. Some tomato pulp or chutney is very good. Mix lightly so as not to make it pasty. Remove from fire, add a beaten egg, and press into a plain buttered mould. Tie down with buttered paper and steam for one hour. Turn out and serve with tomato sauce. It may also be garnished with slices of hard-boiled egg, beetroot, fried tomatoes, &c.

Kedgeree.

A very good kedgeree is made with much the same ingredients as above. The lentils may be left out, and chopped tomato or carrot flaked (on one of those threesome graters is best) and fried along with the onion, may be used instead. The rice must be boiled as for curry and made very dry. Boil 2 or 3 eggs hard, chop finely, and mix with the other ingredients in saucepan. Make all very hot, and serve piled up on hot dish with any suitable garnish and curry or tomato sauce. A spoonful finely chopped parsley would be an improvement to both this and rice mould. Fried parley and thin slices of lemon make a suitable garnish for this and similar dishes, while parsley fried in fat at a low temperature, 200 degrees, crushed and sprinkled over a mould, cutlets, &c., both looks and tastes good. Any kedgeree that is left over will make excellent cutlets for breakfast, &c.

Macaroni Mould

is made by using cooked macaroni instead of rice in recipe for rice mould.

Macaroni Timbale.

Boil 6 ozs. long pipe macaroni — in as long pieces as convenient — in salted boiling water 20 to 25 minutes, and drain. Have a plain mould — a small enamel pudding basin is best — butter it well, and line closely round it with the macaroni. Fill in with any savoury mixture, such as lentils, tomatoes, mushrooms, celery, carrots, &c. Put more strips of macaroni or a slice of buttered bread on the top. Cover with buttered paper and steam 1-1/2 hours. Turn out and serve with sauce. Garnish suitably, cooked tomatoes, &c.

Roman Pie

Boil 4 ozs. macaroni and drain. Butter a pie-dish and put in half the macaroni. Scald 4 or 5 tomatoes in boiling water for a few minutes, when the skin will come off easily. Boil 2 eggs hard and slice. Have 2 ozs. cheese grated, and sprinkle half of it over the macaroni, then put half of the eggs and half the tomatoes. Season with salt, pepper, and a little grated onion (I keep an old grater for the purpose). Take 8 or 10 medium-sized flap mushrooms, if to be had, clean and trim, removing the stalks. Add a layer of them, and repeat as before, but put the mushrooms before the tomatoes. Cover the top thickly with bread-crumbs. Make a stock with the trimmings of mushrooms and tomatoes. Put dessertspoonful butter in saucepan, stir in *half* teaspoon flour, same of made mustard, and perhaps a little ketchup. Add the stock — there should be about a teacupful — stir till it boils, and pour equally over the pie. Dot over with bits of butter, and bake one hour in fairly brisk oven.

In case this pie may be voted rather elaborate by some, I may say that it is excellent with several of the items left out. The eggs, mushrooms, cheese—any one of these, or all three may be dispensed with, and what may be lost in richness and flavour will be compensated in delicacy and digestibility. Any of this pie that is left over may be made into cutlets, so that one can have a second dish with little extra trouble.

NOTE.—When fresh tomatoes are not to be had tinned ones will do.

Tomato and Rice Pie.

Wash well a teacupful good rice—Patna is best for this dish as it does not become so pulpy as the Carolina—and put on with cold water to cover and a little salt. Allow to cook slowly till it has absorbed all the water. Add a little more if too dry, but do not stir. Peel 1 lb. tomatoes, cut in 1/2 inch slices and put a layer in buttered pie-dish. Put in the rice—or as much of it as wanted—sprinkle with curry and seasoning to taste. Put rest of tomatoes on top, more seasoning, and layer of bread-crumbs. Put plenty of butter on top and bake 3/4 hour.

Note.—Tinned tomatoes may be used when fresh ones are not at hand. Any of the dishes with tomatoes, rice, &c., may have grated cheese or hard-boiled eggs added at discretion, and in this way the several dishes may be varied and adapted to suit different tastes and requirements.

Casserole of Rice.

Wash well 6 ozs. whole rice and drain. Melt in saucepan 2 ozs. butter or 1-1/2 ozs. Nut Butter. Put in rice with as much white stock or water as will cover it, a little salt, pinch mace if liked, and allow to simmer very slowly or steam in double boiler till quite soft. Stir well, and if too stiff add a little more water, but it must not be 'sloppy.' Beat well till quite smooth and set aside to cool. Butter plain mould and line with rice nearly an inch thick. Fill in with any savoury materials, such as tomatoes, mushrooms, onions, celery, fried slices of carrot, lentils, &c. An hour before dinner cover with buttered paper and steam. Turn out on hot dish, cut a round off the top, and either serve as it is with garnish and sauce, or brush over with beaten egg, sprinkle with fine crumbs, and brown in brisk oven.

Vegetable Goose.

Put 2 teacupfuls crumbs in basin and pour boiling water or milk over. Let soak for a little, then press out as much moisture as possible. Add dessertspoonful grated onion, teaspoonful chopped parsley, pinch herbs or mace, salt, white pepper, 1/2 teaspoonful "Extract," and some mushroom ketchup. Mix all well, and add a beaten egg to bind. If too stiff add a little milk, stock, or gravy. Put in flat well-buttered baking-tin, and bake for about an hour, basting occasionally with butter or vegetable fat. Serve with fried tomatoes or any suitable sauce.

Celery Souffle.

This is exceedingly good if nicely made and served. Clean 1/2 lb. white crisp celery and cut small. Simmer in enamel pan or steam with as little milk as possible till tender, then boil rapidly to reduce the liquid. Rub through a sieve and set aside to cool. Beat 1 oz. fresh butter to a cream and add yolks of 2 eggs, one at a time, beating well in, also barely 1 oz. grated cheese and seasoning to taste. Mix well. Beat whites of 3 eggs quite stiff and mix in very lightly. Butter souffle tin and tie band of buttered paper round, to come 2 inches above the rim. Fill in mixture—not more than three-fourths full, and steam very gently in barely an inch of water for 1 hour. Turn out on *very* hot dish and serve immediately, or slip off paper band and pin hot napkin round. If allowed to stand any time it will be quite flat before serving. A rather daintier if more troublesome way is to fill small souffle cases three-fourths full with the above mixture. Sprinkle a little grated Parmesan cheese and celery, salt on the top, and bake in hot oven 10 minutes. Arrange tastefully on hot napkin.

NOTE.—Very dainty souffle cases are now to be had in white fluted fire-proof china. These can come straight to table without any trouble of swathing with napkins, paper collars, and the like.

Celery Cream

is another delicacy well suited to a special occasion. Prepare and cook celery as for souffle, drain and rub through sieve. Have enamelled or earthenware saucepan on the table, rub the bottom with a little butter, and break in 2 large eggs or 3 small ones. Season with white pepper, celery salt, lemon juice, mace, &c., and beat slightly. Take 1/2 gill cream and same of milk, drained from the celery, and add to eggs, &c. Place over a slow fire, or better still, a gas stove turned low, and stir till the mixture thickens, but it must not boil, then add the celery and mix. Have one large timbale mould or 8 to 10 small ones well buttered, fill in with the cream, cover with buttered paper, and steam very gently till set—30 minutes if large mould—10 minutes if small ones. If a large one turn out and fill in centre with tomatoes, mushrooms, &c. If small ones arrange round ashet with baked tomatoes, spinach, green peas, &c., in the centre of the dish.

* * * * * *

A PERFECT NUT FAT!

PURE :: WHITE :: TASTELESS

PREPARED FROM FINEST NUTS ONLY

NUTTENE

Unsurpassed for all kinds of Pastry and Confectionery.

NUTTENE was exclusively used at the Vegetarian Schools, Melrose and
Penmaenmawr, and Vegetarian Restaurants, Dublin and Edinburgh Exhibitions.

Send for complete New Price List, with Recipes, to the Manufacturers:

Chapman's Health Food Stores,
EBERLE STREET, LIVERPOOL.

* * * * *

A FEW OF "PITMAN" 1001 DELICIOUS

HEALTH FOODS

In place of Meat, and Free from its Dangers.

BRAZOSE MEAT.—Made from Brazils. Quite different to all other Nut
Meats. Makes splendid Sandwiches, Sausage Rolls, Savoury Roasts, and Irish
Stews. Per tin—1/2-lb., 10d.; 1-lb., 1/6; 1-1/2-lb., 2/1; sample, 4d.

VIGAR BRAWN.—The Superb Cold Dish. Tomato or Clear. Per mould, 1/-

TOKIO BAKED BEANS, with Tomato and Nut Sauce. Hot or cold, makes a splendid Dinner Dish. Per Jar, 1/-; sample jar, 4d.

BAKED BEANS, with Tomato Sauce. Per tin, 9d.; sample tin, 2d.

CURRIED BEANS, with Savoury Sauce. Per tin, 9d.; sample tin, 2d.

VEGETABLE SOUPS.—In 12 varieties. Per tin, 2-1/2d. Each tin makes a pint. Per doz., 2/6

NUTMARTO POTTED PASTE.—Far superior to Meat and Fish Pastes. Per tin, 3-1/2d.; per glass jar; 5-1/2d.

VIGAR GRAVY ESSENCE.—Delicious flavour. Add but a few drops to water. Per bottle, 6d., 1/-, and 1/6; sample, 2d.

One Penny Packet Health Wafers With Two Ripe Bananas, INSURE a Perfect Meal.

All those interested in Health Foods and Perfect Health should read "PITMAN" Health, from Food Library, No. 1 to 8. One Penny each, post free 1-1/2d.; or the 8 for 10. Full Catalogue of Health Foods, with "Diet Guide," post free one stamp. A wise selection of Health Foods will give you

PERFECT HEALTH

and Digestion, and so enable the system to perform the Maximum amount of Work—both mental and physical—with the Minimum amount of Fatigue.

Ask your Stores for them: or assorted Sample Orders of 5/- value, carriage paid, from the Sole Manufacturers:

"PITMAN" HEALTH FOOD STORES,

155 Aston Brook St., Birmingham.

(The Largest Health Food Dealers in the World.)

* * * * *

Asparagus Cream

is prepared in the same way, putting tender cooked asparagus in place of the celery.

Celery or Asparagus Quenelle

is made much in the same way. To every teacupful celery or asparagus pulp allow 2 cupfuls fine white bread crumbs. Beat up two or three eggs, add, and mix well. Steam in large or small moulds, or divide into spoonfuls, shape round, and poach in boiling water, stock, or milk. Serve with cooked tomatoes or sauce, or they may be put in tureen with clear or white soup.

Many toothsome variants of the foregoing recipes will suggest themselves as one goes along, so that it is needless to detail each at length. Thus, fritters, moulds, quenelles, &c., may be varied at pleasure by substituting cauliflower, the white of spring onions or leeks, &c., for the celery or other ingredients mentioned. By the way, we do not appreciate the food value of leeks as much as we ought. A dozen or so of the thickest

Leeks Stewed or Steamed

in milk or stock, and served with the liquor made into a white sauce, is a dish as delicious as it is wholesome and blood-purifying.

Needless to say, everything should be the best of its kind and absolutely fresh. To ensure this we should make a point of using as far as possible those which are in season at the time, as however well preserved they may be, vegetables, especially the finer sorts, lose in flavour and wholesomeness every hour between the garden and pot.

Substantial Savouries.

We come now to the more substantial savouries which form the staple part of the ordinary family dinner. These, along with soup and pudding, will furnish an excellent three-course meal, and where time—or appetite—is limited, as in the rush to and from school or business, two sources will be found ample.

German Lentil Stew.

Among the various pulse foods, of which there are fifty or sixty different kinds, though only some half-dozen are at all well-known, German lentils are one of the most valuable. In this country they are but little used, but they only need be known to be heartily appreciated. As far as my experience goes, every one who has once sampled them is loud in their praises. Even in those households where meat is used they might come as a change and variety, and help to solve the problem of that typical, much-to-be-pitied housekeeper who so pathetically wished there might be "a new animal" discovered!

Well, "to return to our" — ahem — lentils. These German or Prussian lentils are quite different from the ordinary yellow kind. They are green or olive coloured, much larger, and of a flat tabloid shape. They are exceedingly savoury, and — if that is any recommendation — so "meaty" in flavour that it is almost impossible to convince people that they are quite innocent in that respect. They are usually sold at about double the price of yellow lentils, and even then are very cheap; but this is a fancy price, charged because of their being a novelty, and I may say that I get the very finest quality, perfectly clean and free from grit, at the extremely low price of 2d. per lb.

To make a stew, which is the basis of a number of other dishes, take 1/2 lb. German lentils and scald for a minute or two in boiling water to make sure that they are thoroughly clean. Drain, and put in good-sized saucepan with plenty of fresh boiling water, and allow to simmer *very gently* for an hour. In another stewpan melt 1 oz. butter, and into that shred very finely two or three onions. Cover, and cook 10 to 15 minutes to bring out the flavour. They may brown or not as preferred, but there must not be the least suspicion of burning. Turn the lentils into this pan, add some chopped celery if at hand — it is very good without, but to my taste most dishes are improved by celery — and allow to simmer an hour longer. See that there is plenty of water — there should be a rich brown gravy. Add seasoning to taste of salt, pepper, Jamaica pepper, parsley, &c. A few tomatoes may be added, or carrots, turnips, &c. A few ozs. macaroni, par-boiled in salted boiling water and added an hour or less before, will make one of the many pleasing varieties of this dish. Serve like a mince, garnished with sippets of toast or fried bread, or toasted Triscuits.

Savoury Pot-Pie.

Line a pudding basin with suet paste [Footnote: See pastry.], and fill in with lentils cooked as above, and tomatoes, or any vegetables, such as carrots, turnips, cauliflower, beetroot, &c., to keep the mixture from being too heavy, for whatever may be thought to the contrary, there is a much larger proportion of solid nutriment to the bulk in pulse foods than in the "too, too solid flesh" which we esteem so highly. And, at the risk of wearying readers with reiteration, I must say again that herein lies the danger. Quite a number of people have told me that they would like such foods, but *they* could not take enough to keep up their strength, and were reproachfully incredulous when, ignoring the gentle insinuation as to *other* people's capacity, I told them the great difficulty was to take little enough! But we must finish the pot-pie. Put a thin round of paste on the top. Wet the edges and press together, tie down with greased paper, and steam 2 to 3 hours. Turn out and send to table with suitable hot garnish.

The same paste may be made into little balls or flat cakes and put to cook with lentil stew, but great care must be taken to see that there is plenty gravy, and that they cook very gently, for if they "catch" ever so slightly they are spoiled. All danger of this can be avoided by steaming in a basin or jar instead of cooking in open pan.

Savoury Brick.

Take about 2 teacupfuls cooked German lentils—not too moist. Put in a basin and add a cupful fine bread crumbs, and a cupful cold boiled rice or about half as much mashed potatoes. Add any extra seasoning—a little ketchup, Worcester sauce, Marmite or Carnos Extract, &c.—also a spoonful of melted butter. Mix well with a fork and bind with one or two beaten eggs, reserving a little for brushing. Shape into a brick or oval, and press together as firmly as possible. Brush over with beaten egg, put in buttered tin, and bake for half-an-hour. Or it may be put in saucepan with 1 oz. butter or Nut Butter that has been made very hot. Cover and braize for 10 minutes. Turn and cook for another 10 minutes. Add a little flour and seasoning to the butter, and then a cupful boiling water, stock, or diluted "Extract," and allow to simmer a little longer. Serve with garnish of beetroot or tomatoes.

This can also be made into a delicious

Cold Savoury.

Bake or braize as above. Remove to the ashet on which it is to be served.

Allow to get quite cold, then glaze. [Footnote: See Glaze.]

Sausages

are made of the same ingredients as savoury brick. Pound well in a basin, so as to have all the materials nicely blended, or put in a saucepan over gentle heat, and mash well with a wooden spoon. See that the seasoning is right. Some chopped tomatoes and mushrooms are an improvement, also some grated onion, ketchup, and "Extract." These should be put in saucepan with a little butter until lightly cooked, then the lentils, &c., should be added, the whole well mixed and turned out to cool. When quite cold, flour the hands and form into small sausages. Brush over with beaten egg and fry, or put on greased baking tin and bake till a crisp brown. They may need a little basting, or to be turned over to brown equally.

The filling for

Sausage Rolls

is compounded exactly as above, but should be rather moister, and have more butter added to prevent their being too dry. Have quantity required of rough puff pastry. [Footnote: See Pastry.] Roll out and divide into 9 or 10 4-inch squares. Put a little sausage meat in centre, wet the edges and fold over. Press the edges lightly together with pastry cutter, if you have one, brush all over with beaten egg except the edges. Place on oven plate and put at once in hot oven. Bake 20 to 30 minutes. They may be served either hot or cold, but are best hot. They can easily be re-heated in oven at any time.

Fifeshire Bridies

may have the same filling put in plain short crust, or raised pie-crust, rolled very thin and cut in oval or diamond shapes. Fold over, and turn up the under edge all round. Brush over with egg and bake — if raised pie crust — in rather a slower oven.

Rissoles.

Roll out rough puff or short crust very thin, stamp out into rounds, put a little of the mince on one, wet edges and put another on top, press very firmly together, brush over with egg and fry in deep, smoking-hot fat.

German Pie.

Take an ordinary pie-dish, such as used for steak pie. Have one or two large Spanish onions half-cooked, remove the centres, and put in pie-dish. This will serve both to keep up the paste and to hold gravy. Fill up the dish with partially stewed German lentils, and either sliced tomatoes or pieces of carrot and turnip first fried in a little butter. There should also be plenty of chopped onions put in the bottom of the dish, which should be buttered. Fill nearly up with well-seasoned stock, "Extract," gravy, or water, cover with rough puff paste, and bake for an hour or longer, according to size. There should be a hole in top of pastry, covered with an ornament, which could be lifted off, and some more gravy put in with a funnel. Serve very hot. If to be used cold, a little soaked tapioca should be cooked with it, or some vegetable gelatine might be dissolved in the gravy.

By way of variety, a few force-meat balls may be put in; also mushrooms when in season.

Haricot Pie

is made much the same as above, substituting Butter Beans or Giant Haricots for the German lentils. They should be soaked all night and nearly cooked before using. Put in a layer of beans, sprinkle in a little tapioca, then put a layer of sliced tomatoes and repeat. Fried beetroot may be used instead of tomatoes, and crushed vermicelli or bread crumbs instead of tapioca.

Haricot Raised Pie,

which is very good to eat cold for pic-nic luncheon, &c., is made as follows:—Soak 1/2 lb. large beans all night, when the skins should come off easily, and put to stew or steam with butter, shred onions, and a very little stock or water till soft, but not broken down. Set aside to cool. Prepare a raised pie case [Footnote: See Pastry.], put in half the beans, a layer of sliced tomatoes, and a layer of hard-boiled eggs. Repeat. Put on lid, which should have hole in centre, and bake in a good, steady oven for an hour. Meanwhile, have some strips of vegetable gelatine soaking, pour off the water, and bring to boil in a cupful well-seasoned stock, "Extract," gravy, &c. Stir till gelatine is dissolved, and when the pie is removed from the oven, pour this in. When cold this should be a firm jelly, and the pie will cut in slices. If tomato or aspic jelly is prepared, some of that would save trouble. Melt and pour in.

There are many other toothsome ways of serving haricot and butter beans. In every case they should first be well washed, soaked, and three-parts cooked with stock or water, butter, onions, and seasoning.

Savoury Haricot Pie.

This is made without paste. Put a layer of beans in buttered pie-dish, then pieces of carrot and turnip—previously par-boiled—to fill up the dish. Pour in a little gravy. Cover with a good white sauce, well seasoned with made mustard, chopped parsley, &c., and coat thickly with bread crumbs. Dot over with bits of butter, and bake 30 or 40 minutes.

Many variations will suggest themselves—cauliflower, parsnips, vegetable marrow, sliced tomatoes, beetroot, &c., instead of the other vegetables. Or the same ingredients as in the first haricot pie might be used, with the crumbs instead of pastry.

Haricot Ragout.

Half pound soaked beans boiled till tender in one pint water, with butter and sliced onions. Drain, but keep the liquor. Slice some carrots and turnips thin, fry lightly, and then simmer in the liquor for half-an-hour. Put a little butter in stewpan, slice and cook two onions in that, with the lid on, stir in a tablespoonful flour, and add the haricots, vegetables, and the liquor. Simmer gently till all are quite cooked, and serve. Some tomatoes or a little extract may be added, and it can be varied in many other ways.

Golden Marbles.

Take nearly a teacupful of haricots pulped through a sieve, and add to this 2 ozs. bread crumbs. Same of mashed potatoes; a shallot finely minced, or a spoonful of grated onion. Beat up an egg and add, reserving a little. Mix thoroughly, and form into marbles. Coat with the egg, toss in fine crumbs, and fry in smoking-hot fat till golden brown in colour.

Haricot Kromeskies

can be made with the same mixture as for marbles. Some chopped tomatoes, beetroot, or mushrooms may be added. If the mixture is too moist add a few more crumbs; if too dry add a little ketchup, milk, tomato juice, &c. Form into sausage-shaped pieces or small flat cakes. Dip into frying batter, and drop into smoking-hot fat. When a golden brown lift out, and drain on absorbent paper. Serve them, as also the golden marbles, on sippets of toast or fried bread with tomato or parsley sauce.

Haricot Croquettes or Cutlets

are of course made with any of these mixtures. Shape into cutlets, egg, crumb, and fry in the usual way.

There are an immense number more dishes which can be made with pulse foods, for which I have not space here. There are also a number of new varieties of pulses being put upon the market, which can be used with advantage to vary the bill of fare and enlarge its scope.

Giant Split Peas

are especially good, and might be used in any of the foregoing recipes in place of haricots. One advantage is that they do not require soaking. If scalded with boiling water, drained, and put to cook in fresh boiling water, they will be quite soft in little over an hour.

The best quality of butter beans also need no soaking. After scalding for a few minutes the skins come off quite easily. There is also a new variety called

Butter Peas, or "Midget" Butter Beans,

which I can heartily recommend. In appearance they resemble the small haricots, but are much finer and boil down very quickly. They make a very rich white soup, and may, of course, be used for any of the savouries for which recipes are given. Scald with boiling water (or they may merely be rubbed in a clean coarse cloth), plunge into more boiling water—the quantity proportioned to the purpose for which intended, soups, stews, &c.—and simmer till just tender, but not broken down.

Though they can be made up in a host of ways they are perhaps nicest as a simple stew. When just cooked—and great care must be taken not to _over_cook, for much of the substance, as well as the delicacy of flavour, is lost if we do—have a saucepan with some shred onions, sweated till tender, but not in the least coloured, in a little butter. Stir in a spoonful of flour, and when smooth a gill of milk, or the stock from the butter peas. Stir till it thickens and add the peas themselves, and any extra seasoning required. See that all is quite hot, and serve garnished with sippets of toast.

Brown Lentils

also furnish us with unlimited possibilities for new dishes. They are as yet rather difficult to procure, but need only to be known to become very popular. They somewhat resemble German lentils, but are much browner and smaller. Being so small, extra trouble must be taken to see that they are clean and free from grit. They can be used in place of German lentils for any of the soups or savouries for which recipes are given. They cook very quickly, and care must be taken with them also not to waste any of their goodness up the chimney.

Toad-in-a-Hole.

Make the sausages the same as in previous recipe, only using brown lentils instead of German lentils. Put in a buttered pie-dish and pour over the following

Batter.

Beat up one or two eggs. Add 3 tablespoonfuls flour, and by degrees two gills milk, also seasoning of grated onion, chopped parsley, white pepper, "Extract," &c. While

Fresh Green Peas or Beans

are to be had, one need not be confined to the dried pulses. Cook the peas, broad beans, or French beans, as directed in "Vegetables." Serve with poached or buttered eggs, fried or baked tomatoes, &c.

One might go on *ad infinitum* to suggest further combinations and variations of the different pulse foods, but these must be left to suggest themselves, for I must now pass on to another class of foods.

NUT FOODS.

We are only beginning very slowly to recognise the valuable properties of nuts and their possibilities in the cuisine. Indeed, there is a rather deep-rooted prejudice against them as food, people having been so long accustomed to regard them as an unconsidered trifle to accompany the wine after a big dinner, and as in this connection they usually call up visions of dyspepsia, many people regard the idea of their bulking at all largely in a meal with undisguised horror. I remember a lady saying to me that she was quite sure a meal composed to any extent of nuts would *kill* her, for if she took even one walnut after dinner it gave her such pain. That rather reminds one of the story of a half-witted fellow who used to go about the country doing odd jobs, and asking in return a meal and a shake-down of straw or hay.

He always expressed astonishment at folks being able to sleep on feather beds, his aversion being founded on the fact that he had one night lain down on the hard ground with a single feather under him. "An' if I had sic a sarkfu' o' sair banes wi *ae* feather," he argued, "what like maun it be wi' a hale bed?"

Well, I can assure readers that whatever may be the troubles of a solitary nut in an oasis of good things, it is very different when nuts are taken alone or in a suitable and simple combination. Most of our digestive troubles are due to an excess of proteid matter, which clogs up the system, and either lodges in the body in the shape of some morbid secretion, or tries to force its way out in an abnormal way, as by the skin. Now, nuts are very rich in proteid, or flesh-forming matter, and it stands to reason, that if we superimpose them on an already full, or overfull, meal, the result is surfeit, and however wholesome or digestible this excess matter may be in itself, it may, and usually does, work harm in more or less obvious ways.

But curiously enough, this does not always work out with mathematical directness. Most things in the physical, as in the metaphysical, world work out as Ruskin says "not mathematically, but chemically." Though this may seem a far-fetched simile in connection with our dinner, it is a true one. To get back to our nuts. If after a meal of several courses, rich in quality and variety, highly-spiced and flavoured, and perhaps interspersed with little piquant relishes, serving to whet the appetite for the next course, one takes only a very few nuts, or an apple, or a banana, the probability is that "these last" will give the most direct trouble. The gastric juices may be already exhausted, and the nuts, therefore, lie a hard undigested mass on the stomach; or the apple digesting very quickly, and being ready to leave the stomach some hours before its other contents, but having to bide their time, ferments and gives off objectionable gases. Thus, the innocent fruit gets the blame, and the fish, game, or meat go free. Another way in which fruits may prove indigestible, through no fault of their own, is because of the unsuitable combination in which they are eaten. Most nuts, with the exception of chestnuts, which are largely composed of starch, consist almost entirely of fat, which, unless it meets with an alkali to dissolve it, is digested with great difficulty. The uric acid in flesh tends to harden this fat, and so retards digestion.

The medical faculty now recognise the nutritive properties of nuts, as also their wholesomeness and freedom from all toxic elements, and at all sanatoria for the treatment of rheumatic and gouty affections a nut and fruit diet is the established regime. We need not, however, go to an expensive sanatorium to enjoy this food, but may cure, or better, prevent these diseases in our own homes.

They are, I believe, best in their natural state, along with fresh fruits, salads, and the like, but there are also many dainty dishes, in the composition of which they may be used with advantage.

Mock Chicken Cutlets

only require to be known to be appreciated. Grate 1/4 lb. shelled walnuts—this is best and easiest done by running through a nut-mill, but these are not expensive, as they may be had from 1s. 6d.— or Brazil nuts, and add to them two teacupfuls bread crumbs, mix in 1/2 oz. butter, spoonful onion juice, and a little mace, white pepper, salt or celery salt. Melt 1/2 oz. butter in saucepan. Mix in a teaspoonful flour, and add by degrees a gill of milk. When it thickens add the other ingredients. Mix well over the fire. Remove and stir in a beaten egg and teaspoonful lemon juice. Mix all thoroughly and turn out to cool. Form into cutlets, egg, crumb, and fry. Serve with bread sauce or tomato sauce.

Brazil Omelet.

Take 2 ozs. shelled Brazil nuts and rub off the brown skin. If they are put in slow oven for 10 minutes, both shell and skin will come off easily. Flake in a nut-mill or pound quite smooth. Add the yolk of hard boiled egg, a teaspoonful ground almonds, or almond meal, and make into a paste. Then add some grated onion, a tablespoonful baked or mashed potato, the same of bread crumbs, and seasoning to taste. Mix well, and add the yolks of two eggs beaten up, and after mixing thoroughly, stir in lightly the two whites beaten quite stiff, butter a shallow tin or soup-plate, and pour in the mixture. Cover and bake gently, till set—about an hour. When cool, cut into neat shapes, egg, crumb, and fry. The same mixture will also make a delicious

Brazil Souffle.

Add another white of egg stiffly beaten, and steam gently for 30 minutes.

Brazilian Quenelles.

Add another two tablespoonfuls bread crumbs, and leave out the potato; use three eggs, but beat yolks and whites together. Butter one large or a number of small moulds, fill with the mixture, and steam gently for 20 to 40 minutes, according to size; turn out, and serve, if large, with slices of tomatoes baked or fried, arranged round. If small ones, have tomatoes piled up in centre and quenelles placed round.

A number of other savouries, in which nuts form a part, can be made by substituting grated walnuts, Brazil nuts, almonds, almond meal, Barcelonas, &c., for peas, beans, lentils, &c., in the previous recipes. As they are highly nutritive and concentrated, they must be used sparingly, however, along with plenty of bread crumbs, rice, and the like. There is no need to detail these, but I will give one to show what I mean.

Walnut Pie.

Run 4 ozs. shelled walnuts through the nut-mill—this will give about a teacupful. Have some whole rice boiled as for curry, and put a layer of that in buttered pudding dish. Put half of the grated nuts evenly on the top, then a layer of tomatoes seasoned with grated onion, parsley, salt, pepper, pinch mace, ketchup, &c. Repeat. Cover thickly with bread crumbs, pour some melted butter over and bake till a nice brown. If rather dry, pour some tomato sauce, diluted extract, gravy, &c., over. Serve with tomato or other sauce.

The same ingredients may be put in a buttered mould and steamed, or the whole may be mixed together, a beaten egg added, then made into one large or a number of small rolls, place in baking tin, put some butter on the top and bake, basting and turning now and then.

Prepared Nut Meats.

Of late years since the food value of nuts has been recognised, the attention of specialists has been turned in their direction with very practical results. Quite a number of excellent "Nut Meats" are now upon the market, and each year adds to their variety, so that one's storeroom can be supplied in a way that was impossible only a few years ago. For a cold luncheon dish Mapleton's Fibrose, Almond Nut Meat, and

Savoury Nut Meat

Are very good. The latter is put up in air-tight glass dishes. Tomatoes or any vegetable may be served with it. Then Meatose, Nut-Meatose, Vejola, Nutvego, &c., are all excellent. The

"F.R." Meatose

Is specially fine. These "Meats" are all ready for use, and may be made up in any of the ordinary recipes for Stews, Pies, Sausage Rolls, &c. One dish which most people would like is

Curried Nut Meat.

Melt 1 oz. butter in stewpan, and into that put a tablespoonful finely shred or grated onion, a few slices of tart apple or a little rhubarb, and, if possible, some tomatoes—fresh ones peeled and sliced are best, but the tinned ones will do very well. Stir in a dessertspoonful flour and curry powder to taste, and pour on boiling water, stock, or gravy as required. Slice the nut meat and lay it in. Cover, and cook gently for about half an hour. Serve with plain boiled rice.

I have not space to give further recipes, but would just add a word of caution—use very sparingly. They are highly concentrated and nutritious foods, and a large quantity is not only unnecessary, but harmful.

In addition to above, there are the products of the International Health Association, "the pioneer manufacturers of health foods," who have within the past year removed their works into the country (Stanborough Park, Watford, Herts). Then Messrs Winter, Birmingham, "Pitman," Birmingham, and Messrs Chapman, Liverpool, have a number of excellent nut meats, fuller reference and recipes for which will be found in the chapter on "Health Food Specialties" at end of book.

CHEESE SAVOURIES.

Many excellent cheese dishes, such as macaroni cheese, &c., are to be found in the category of every household, so it will be needless to detail those which are most generally known. Cheese is highly nutritious, and not indigestible for those in ordinary health, if taken in moderation and combined with other lighter and bulkier foods. Cheese with rice, bread crumbs, macaroni, tomatoes, &c., is exceedingly good. It should be used very sparingly, or not at all, in dishes which contain pulse, nuts, or eggs. It should always be grated so that it can be mixed thoroughly with the other ingredients.

Rice and Cheese.

Half teacupful rice, 2 ozs. grated cheese, one egg. Wash rice and put on with cold water to barely cover, and pinch salt. When that is absorbed, add milk enough to swell and cook the rice thoroughly without making it sloppy. Remove from the fire and stir in the cheese, seasoning of salt, pepper, or made mustard, pinch cayenne, and the egg beaten up. Turn into buttered baking dish and bake gently till set and of a pale brown—cheese dishes must never be done in too hasty an oven, as they acquire an unpleasant flavour if in the least burnt. Turn out on hot ashet, and serve garnished with slices of hard-boiled egg or fried tomatoes.

Cheese and Semolina.

Four ozs. cheese, breakfast cup milk, 1 oz. semolina, 2 eggs. Bring milk to boil and stir in semolina. Cook till it thickens; remove from fire and stir in the cheese, pinch cayenne, and yolks of eggs beaten up, beat up whites stiffly, and mix in lightly. Turn into buttered pudding-dish and bake gently till ready—about half-an-hour. This mixture, and the previous one, may also be steamed for about 40 minutes. Serve with fried tomatoes or tomato sauce.

I may say here that tomatoes go very well with cheese in almost any form. A nice variety of rice and cheese can be contrived as follows:—Put half of the cooked rice in pudding dish, put breakfastcupful tomatoes in saucepan with a little butter, the cheese and seasoning, and just stir over the fire till quite mixed. Put half over the rice, then the rest of the rice, and the other half of the tomato mixture. Coat thickly with crumbs, put some butter on top, and bake.

Cheese Souffle.

Two tablespoonfuls grated cheese, 2 eggs, 1-1/2 gills milk. Beat yolks of eggs and mix in cheese, milk, pepper, salt, pinch cayenne, and, lastly, the whites beaten quite stiff. Make souffle tin very hot, pour in mixture, and bake in quick oven till set—15 to 20 minutes. Serve very hot.

Scotch Woodcock.

This is a favourite savoury in many non-vegetarian households. There are numerous different recipes, which will doubtless be well known, but the following is quite new and original. Prepare some slices of buttered toast or fried bread, take about 1 lb. fresh tomatoes or a large cupful tinned ones drained from the liquor, put in saucepan with a little butter and grated onion, and stew gently till the tomatoes are pulped. If at all stringy, put through a sieve. Add 2 ozs. grated cheese, seasoning to taste, and stir over gentle heat till quite thick. Spread a layer of this mixture on each slice of toast and pile on the top of each other. Reserve a little of the mixture and to it add some tomato juice or milk, mushroom ketchup, or diluted extract. Make very hot and pour right over, sprinkle with chopped parsley, and garnish with slices of hard-boiled eggs—or these might have the whites chopped up and the yolks grated over the top. Serve very hot. A very tasteful effect is made by having the slices of toast, which may be round, oblong, &c., graduating pyramid-wise from a large one at the bottom to a small one at the top.

Cheese Straws (1).

Rub 2 ozs. butter into 4 ozs. flour. Add 2 ozs. grated cheese, a little mustard and cayenne, and make into a stiff paste, with the yolks of 2 eggs or one whole egg beaten up. Roll out thin, cut into straws, lift on to baking sheet carefully with a knife, placing them a little apart, and bake a pale brown—about 10 minutes in moderate oven. Another way is to break off small bits of the paste and roll into thin pipes on a floured board. Savoury

Cheese Biscuits

are made by cutting above paste, rolled very thin, into oblong or diamond shapes, with pastry cutter. Bake in same way. Serve either hot or cold. Spread with a little Marmite and savoury tomato mixture, or sandwich this between two biscuits.

Cheese Straws (2).

Two ozs. cheese, same of batter, flour and fine white crumbs. Add seasoning, and make into paste with one egg, roll out, stamp out a few rings, make the rest into straws, bake and put a bundle of straws into each ring.

Parmesan Puff Pie.

Prepare some cheese pastry, as for "Straws No. 1," and with it line a round shallow tin or tart ring. Common short or puff pastry will do, but the cheese pastry is nicer. Fill in with rice or crusts to keep in place. Bake rather briskly, and remove from the tin. Fill in with the following mixture:—In a saucepan melt 1 oz. butter, and into that stir 1 oz. flour and 1 oz. flaked or ground rice. Add gradually a teacupful milk, and when it thickens, 2 ozs. grated cheese and seasoning, cayenne, and made mustard. Pour into pastry case. Sprinkle a few browned crumbs or shredded wheat biscuit crumbs on the top. Dot over with bits of butter, and bake in moderate oven for about 20 minutes. Put a little more grated cheese on the top and serve very hot.

Small Cheese Tartlets

can be made by dividing same ingredients into a number of small cases or patty tins. Ten minutes should be long enough to bake. Another very good filling for these or the previous puff pie is the mixture given in recipe for Scotch woodcock, while a novel and delicious

Welsh Rarebit

could be made with either of these mixtures, with perhaps a rather more liberal supply of cheese and made mustard spread between slices of hot buttered toast.

Mock Crab

is made with somewhat similar filling, but is best with fresh tomatoes. Remove skin and seeds from 1/2 lb. firm, ripe tomatoes, and cut small; grate 4 ozs. rich, well-flavoured Cheddar cheese. Add to tomatoes in basin with teaspoonful made mustard, yolks of 3 hard-boiled eggs, large spoonful mushroom ketchup, a little extract, and a very little curry powder or paste. Pound all together with back of a wooden spoon till quite smooth. Serve in scallop shells, garnished with the white of egg.

These cheese tartlets, mock crab, patties, &c., can be most acceptably varied by using

Shredded Wheat Biscuits

in place of pastry cases or scallop shells. Use any of the cheese mixtures given for Scotch woodcock, mock crab, &c. With a sharp-pointed knife split the biscuit open and place in buttered tin, with a bit of butter on the top of each, in hot oven till crisp and brown. Remove to hot dish, fill in each biscuit with the mixture made very hot, and pile up more on the top.

Dresden Patties.

Stamp out 6 or 8 rounds of bread, dip quickly in milk, gravy, or diluted extract, and drain—on no account allow to soak. Brush over with egg, toss in fine crumbs and fry. Drain and keep very hot. Prepare a cheese and tomato mixture same as for "Scotch Woodcock," and while in saucepan add 1 or 2 hard-boiled eggs—the white chopped in small dice or tiny strips. Mix lightly over the fire and pile up on centre of each round. Serve on hot napkin, garnished with fried parsley. These patties may also be made with shredded wheat biscuits.

* * * * *

'HYGIENIC TREATMENT'

READERS say it "Beats Beecham's Pills!" and is "Worth its weight in gold!!"

LONDON PUBLISHER says "It ought to be half-a-crown!!!"

For all who are Tired of Drugs and want NO MORE VACCINATION, this is the
Best Book in print.

6d only from your Bookseller, and 9d. post free from

A. S. HUNTER, Zetland House, BRIDGE OF ALLAN

* * * * *

MISCELLANEOUS SAVOURIES.

Scotch Haggis.

"Fair fa' yer honest, sonsy face,
Great chieftain o' the puddin' race."

It is to be hoped the shade of Burns will forbear to haunt those who have the temerity to appropriate the sacred name of Haggis for anything innocent of the time-honoured liver and lights which were the *sine qua non* of the great chieftain. But in Burns' time people were not haunted by the horrors of trichinae, measly affections, &c., &c. (one must not be too brutally plain spoken, even in what they are avoiding), as we are now, so perhaps this practical age may risk the shade rather than the substance.

For a medium-sized haggis, then, toast a breakfastcupful oatmeal in front of the fire, or in the oven till brown and crisp, but not burnt. Have the same quantity of cooked brown or German lentils, and a half-teacupful onions, chopped up and browned in a little butter. Mix all together and add 4 ozs. chopped vegetable suet, and seasoning necessary of ketchup, black and Jamaica popper.

It should be fairly moist; if too dry add a little stock, gravy, or extract. Turn into greased basin and steam at least 3 hours. An almost too realistic imitation of "liver" is contrived by substituting chopped mushrooms for the lentils. It may also be varied by using crushed shredded wheat biscuit crumbs in place of the oatmeal. Any "remains" will be found very toothsome, if sliced when cold, and toasted or fried.

Rolled Oats Savoury.

Put a teacupful Scotch rolled oats in a basin, and pour over 2 cupfuls milk in which some onion has been boiled. Allow to soak for an hour, remove onion, add pinch salt, &c., and a beaten egg. Steam in small greased basin for an hour. May be served with a puree of tomatoes.

Irish Stew.

Pare and slice 2 lbs. potatoes, and about 1/2 lb. each carrots, turnips, and onions. Fry all, except the potatoes, a nice brown in a little butter or fat. Put in layers in saucepan with 2 ozs. fat, salt, pepper, and good stock to barely cover. Simmer very gently for about 2 hours. It may also be baked in pie-dish.

This may be varied in many ways, as by adding layers of forcemeat, pressed lentils, &c. Then there are the various nut meats — Meatose, Vejola, Savoury Nut Meat, &c. — which can be used to great advantage in such a stew.

Scotch Stew.

This is a most substantial and excellent dish. Wash well 1/4 lb. *pot* barley — the unpearled if it can be procured — simmer gently in 1 pint white stock for an hour, then add some carrots, scraped — and if large, sliced lengthwise — 2 or 3 small turnips cut in halves or quarters, or part of a large one in slices, a Spanish onion sliced, or a few shallots, some green peas, French beans, or other vegetables that may be in season; some cauliflower in sprigs is a welcome addition. It or green peas should not be added till 1/2 hour before serving. Simmer till all the vegetables are just cooked, adding more stock if necessary. Serve with a border of boiled pasties, potato balls, or chips.

Poor Man's Pie.

Pare and slice 2 to 3 lbs. potatoes. Slice 1 lb. onions; put half the potatoes in pie-dish, then the onions, and sprinkle over 2 tablespoonfuls tapioca and a little powdered herbs or parsley. Add the rest of the potatoes, dust with pepper and salt, pour in water or stock to within 1/2 inch from top. Put 2 oz. butter or nut butter on the top, and bake in moderate oven about 2 hours.

Vegetable Roast Duck.

Take a good-sized vegetable marrow, pare thinly and remove a small wedge-shaped piece from the side. Scoop out the seeds and water, fill in with good forcemeat, replace the wedge, brush all over with beaten egg. Coat with crumbs, put some butter over, and bake till a nice brown, basting frequently. Serve with fried tomatoes.

An ordinary forcemeat of crumbs, onion, parsley, egg, &c., will do, or any of the sausage mixtures given previously.

Esau's Pottage.

The following I have had given me as the original recipe for "Esau's pottage," but I think it must be more elaborate than that set before the hungry hunter.

One pint lentils and 2 quarts water boiled 2-1/2 hours, then add 1/2 lb. onions, 2 lbs. tomatoes, a little thyme and parsley. Cook all together 3/4 hour longer and add 3 oz. butter and 1 oz. grated cheese just before serving.

Dahl.

Wash well 1/2 lb. rice and allow to swell and soften in just as much water or stock as it will absorb. Cook 1/2 lb. red lentils with stock or water, some grated onion, pinch herbs, little curry powder, and any other seasoning to taste. Make a border of the rice, pile the lentils high in the centre, and garnish with slices of hard-boiled eggs. The lentils are best steamed, as they can thus be thoroughly cooked without becoming mushy or burnt.

Mushroom and Tomato Pie.

For a fair-sized pie get 3/4 lb. medium-sized flap mushrooms, the meadow ones are best, and 1 lb. good firm tomatoes, remove the stalks from the mushrooms and wipe with a piece of clean flannel dipped in oatmeal or salt. Unless very dirty, it is best not to wash them, as that somewhat spoils the flavour. Pare and put a layer in pie-dish, along with slices of tomato, pared and free from seeds. Put a little bit of butter on each, dust with salt and pepper, and repeat till the dish is heaped up. Cover with a good, rough puff paste, and bake till the paste is ready, about an hour. No water should be put in, but the trimmings of the mushrooms and tomatoes should be stewed in a little water, and this gravy may be added with a funnel after the pie is ready.

Mushroom and Tomato Patties.

For these we require some richer puff paste. Prepare and trim a small quantity of tomatoes and mushrooms. Cut rather small and cook gently, with a little butter and seasoning, for 10 or 15 minutes. Allow most of the moisture to evaporate in cooking, as this is much better than mixing in flour to absorb it. When the pastry cases are baked, fill in with the mixture. Good either hot or cold. If baked in patty pans, the mixture should be cold before using. Line in the tins with puff paste, half fill, brush edges with egg or water, lay on another round of paste, press edges together and bake.

Vol-au-Vent.

A delicious vol-au-vent is made with exactly the same filling as above.

Mushroom Pie.

Put on stewpan with a piece of "Nutter" or other good vegetable fat. Cut up one large Spanish onion very small, add to fat and brown nicely. Cover with water and stew along with the contents of a tin or bottle of white French mushrooms (including the liquid), also pepper and salt to taste. Stew till the mushrooms are tender, then take out and chop. Dish along with other contents of saucepan, and when cool add a cup of brown bread crumbs, and one beaten egg. Cover with puff paste or short crust and bake. Serve with brown sauce.

Shepherd's Pie.

Mushrooms same as for mushroom pie, but covered with nicely mashed potatoes, adding pepper and salt to the latter. Beat well and cover, stroke with a fork, and brown in the oven.

BREAKFAST DISHES—Porridge.

"The halesome parritch, chief o' Scotia's food."

In these days of tea and white bread it is to be feared that the "halesome parritch" is now very far removed from the honoured place of chief, and it must be more than a coincidence which connects the physical degeneracy of the Scottish working people with the supplanting of the porridge-pot by the tea-pot. Even in rural districts there is a great change in the daily fare, and there too anaemia, dyspepsia, and a host of other ills, quite unknown to older generations, are only too common. Certainly many people have given up porridge because they found it did not suit them—too heavy, heating, &c.—but we must remember that all compounds of oatmeal and water are not porridge, and the fault may lie in its preparation. It is a pity that any one, especially children and growing youths, should be deprived of such valuable nutriment as that supplied by oatmeal, and before giving it up, it should be tried steamed and super-cooked. It is only by steaming that one can have the oatmeal thoroughly cooked and dextrinised, while of a good firm "chewable" consistency, and not only are sloppy foods indigestible, but they give a feeling of satiety in eating, followed later by that of emptiness and craving for food. The custom, too, of taking tea and other foods after porridge is generally harmful.

Now for the method by which many, who have long foresworn porridge, have become able again to relish it, and benefit by it. Make porridge in usual way, that is, have fast boiling water, and into that sprinkle the oatmeal smoothly, putting about *twice* as much oatmeal in proportion to the water as is usual. Boil up for a few minutes, add salt to taste, and turn into a pudding bowl or steamer. Cover closely and put in large pot with about one inch water or in a steam cooker and steam for five to twelve hours. Eat with stewed prunes, figs, &c., or with butter or nut butter—almond cream butter is both delicious and wholesome. A mixture of wheatmeal and oatmeal, or wheatmeal itself, may be found to suit some better than oatmeal alone. I heard recently of a hopeless dyspeptic who recovered health on a diet composed almost entirely of porridge made of three-parts whole wheatmeal to one of oatmeal. I may add that one must be careful to take a much smaller quantity of this firm, super-cooked porridge, as it contains so much more nutriment in proportion to its bulk.

Porridge made with Scotch Rolled Oats also will be found easier of digestion by some than ordinary oatmeal porridge. This also is best steamed and super-cooked.

* * * * *

Health Foods.

Granose. The Ideal "Staff of Life."

A kernel of wheat is acknowledged to constitute a perfect food, and Granose consists of the entire kernels of choice wheat, prepared by unique processes, so as to afford the most digestible food ever prepared.

Granose is equally beneficial from infancy to old age, in good or ill health. It is a royal dainty, and should take a prominent place on every table.

Granose Flakes, 7-1/2d. per packet.
Granose Biscuits, 7-1/2d. "

Protose. The Standard Nut Meat.

Palatable to the taste, resembling chicken in fibre and flavour, but perfectly free from the tissue poisons that abound in animal flesh.

"Chemically it presents the composition of animal tissue, beef or mutton."—*Lancet.*

Protose is prepared from the best grains and nuts, and is perfectly cooked. It tastes good, promotes health and vigour, and imparts great staying power.

Price: —1/2 lb. tin, 8d.; 1 lb., 1/-; 1-1/2 lb., 1/4

Bromose. The Rapid Flesh-Former.

A combination of predigested nuts and cereals. No better food for consumptives, the "the too-thin," and all who desire the best physical condition.

30 Tablets in box, 1/6

Full List of our Health Foods sent post free on application.

For One Shilling we will send you Samples of 12 of our Health Foods, and Cookery Book.

The International Health Association, Ltd.,

Stanborough Park, Watford, Herts.

* * * * *

The name Plasmon distinguishes our preparations of milk-albumen from all other foods.

One Pound of PLASMON contains the entire nourishment of 30 pints of fresh milk.

Most foods are deficient in proteid, which is required to support life.

PLASMON should be added to all foods because it supplies this element.

Foods mixed with PLASMON are therefore more nourishing than any others.

OF ALL GROCERS, CHEMISTS, AND STORES.

* * * * *

FOR HEALTH, STRENGTH, AND ENERGY

[Illustration]

Doctors counsel the regular use of

Shredded Wheat

"Biscuit" and Triscuit

[Illustration]

Because they are ALL-NOURISHING, NATURAL FOODS.

Made in the wonderful Laboratory of the Natural Food Co., Niagara Falls,
N.Y., U.S.A.

SHREDDED WHEAT products give greater surface for the action of the digestive fluids than that given by any other food.

This ensures Perfect Digestion and Freedom from Constipation.

SHREDDED WHEAT BISCUIT (with milk) for Breakfast and Supper, or basis for Sweets. "Triscuit" (with butter, preserves, cheese, &c.) for any meal. The best basis for Savouries and Sandwiches.

Send 1d. stamp for Sample and Illustrated Cook-Book.

SHREDDED WHEAT CO. (C. E. Ingersoll), 70, St George's House, EASTCHEAP,
E.C.

* * * * *

BREAKFAST SAVOURIES.

Most of the rissoles, toasts, &c., given in the earlier part of the book are suited for breakfast dishes, but we may add a few more.

Savoury Omelets.

Separate the whites from the yolks of 3 eggs, or one for each person; beat up the yolks, and add some grated onion, pepper and salt. Beat the whites till very stiff and mix or rather fold in very lightly. Make a small piece of butter very hot in small frying pan, pour in one-third of the mixture, shake over gentle heat till set, easing it round the edges with a knife, fold over and put on very hot napkin. Repeat till all are done and serve very hot. A little hot lemon juice may be squeezed over, or a spoonful of mushroom ketchup will give a nice relish.

Cheese Omelet

is made by mixing in grated cheese—a dessert spoonful for each egg. The onion may be omitted if preferred without. A pinch cayenne and a little made mustard go well with cheese.

Savoury Pancakes.

Take much the same ingredients as above, but beat yolks and whites together, and add one tablespoonful milk, and a level dessert spoonful flour for each egg. Mix all together some time before using. Make a bit of butter hot in very small frying pan, pour in enough batter to just cover, and cook very gently till set, and brown on the under side. Turn and brown on the other side, or hold in front of hot fire or under the gas grill. Roll up and serve very hot. Ketchup and water, or diluted extract, may be used instead of the milk, and some finely minced parsley or pinch herbs is an improvement.

These omelets and pancakes may be varied by adding tomatoes, mushrooms, &c. Cook very lightly and either stir into the mixture before frying, or spread on the top after it is cooked, and fold or roll up. A mixture of tomatoes and mushrooms is especially good.

Mushroom Cutlets.

Remove stalks and skins from 1/2 lb. flap mushrooms. Clean, chop up, and stew gently in a little butter. Melt 1 oz. butter in another saucepan, stir in 1 oz. flour, and add by degrees a teacupful milk, tomato juice, or extract. When smooth add the mushrooms and seasonings. Stir till smooth and thick, and turn out on flat dish to cool. Shape into cutlets, egg, crumb, and fry.

Asparagus, celery, artichokes, and many other vegetables may be used in the composition of omelets, fritters, cutlets, &c.

If for an omelet, only a very small quantity must be used. One tablespoonful of any of the finer cooked vegetables is enough in proportion to two eggs. When a more substantial dish is wanted, it should take the shape of cutlets or fritters.

Bread Fritters.

Put 6 ozs. fine bread crumbs in a basin and pour over 3 teacupfuls boiling milk. Allow to stand for some time, then add seasoning to taste—grated onion, parsley, ketchup, extract, &c.—and 2 beaten eggs, reserving a little of the white for brushing. Mix and pour into buttered baking tin. Cover and bake in good oven till set—about 1 hour. When cold, cut into nice shapes, brush over with egg, toss in fine crumbs and fry. This may also be served simply baked. In that case, put some bits of butter on top, and bake a nice brown without cover.

Eggs

are, of course, invaluable in many ways besides the more familiar boiled, poached, and scrambled.

Buttered Eggs.

Break number of eggs required in a bowl, melt a nut of butter to each egg in saucepan, pour in the eggs, seasoning, &c., and stir one way over gentle heat till set. About 2 minutes should do. Serve on toast or bread cutlets.

Tomato Eggs.

Have a quantity of tomato pulp made hot in frying pan, and slip in as many eggs as required, gently, so as not to scatter. Allow to poach for about 3 minutes or till the whites are just set. Serve on toast or shredded wheat biscuits. Another way is to cook the tomatoes, and put, with the eggs, on a flat dish, in the oven till set. Serve on same dish, garnished with sippets of toast or toasted triscuits.

Egg Cutlets (Mrs G. D.)

There are many different recipes for these, but the following is an especially good one, for which I am indebted to an Edinburgh friend. Chop very small two firmly boiled eggs, and 2 tablespoonfuls bread crumbs and the same of grated cheese with a pinch of curry powder, pepper, and grated nutmeg. Mix with the yolk of a raw egg. Shape into cutlets, brush over with the white of the egg beaten up a little, toss in fine crumbs, and fry a nice brown. Garnish with fried parsley.

Inverness Eggs.

Boil hard the number of eggs required, remove the shells, and rub each with a little flour. Take a quantity of any of the varieties of sausage meat, for which recipes are given, or a forcemeat, or quenelle mixture will do, add some finely minced parsley, any other seasoning required, and a beaten egg to bind. Mix thoroughly, flour the hands and coat each egg with the mixture, rather less than 1/4 inch thick, and evenly, so that the shape is retained, flour lightly and fry a nice brown. Cut in halves, and serve, round ends up, with tomato sauce.

Toasts

of various kinds come in nicely for breakfast. They can be of ordinary toast, fried bread, or shredded wheat biscuits. The latter are particularly dainty, and may be prepared thus:—Put in buttered baking tin, with plenty of butter on top of each, and place in brisk oven till crisp and brown—about 10 minutes. Pile high with following mixture:—In an enamel frying pan put a teaspoonful butter, and two tablespoonfuls diluted extract or ketchup and water for each egg. When nearly boiling, break in the eggs and stir gently round over a very moderate heat till just set. Season to taste. A little of the sauce made hot might be first poured over the toast or biscuits.

Bread Cutlets.

Have a number of neat pieces of bread about 1/2 inch thick. Dip in milk, gravy, tomato juice, &c., and drain. Do not soak. Brush over with egg or dip in batter, and fry. Serve as they are or with some savoury mince, tomatoes, &c.

Stuffed Tomatoes.

Have number of tomatoes required, equal in size but not too large. With a sharp knife take off a small slice from the stalk end. Scoop out a little of the centre part, mix this with some forcemeat, or sausage mixture, beaten egg, &c., and fill in the cavity. Put some butter on the top and bake. A few chopped mushrooms with crumbs, egg, &c., make a delicious filling.

Cheese Fritters.

Mix 2 tablespoonfuls flour with 1/2 teacupful milk, 2 ozs. grated cheese, teaspoonful made mustard, and the whites of 2 eggs stiffly beaten. Mix well, and drop by small spoonfuls into hot fat. Fry a nice brown and serve very hot.

One might go on indefinitely to detail breakfast dishes, but that is quite unnecessary. It is a good thing, however, to have some simple, easily-prepared food as a regular stand-by from day to day, just as porridge is in some households, and bacon and eggs in others. Variety is very good so far, but we are in danger of making a fetish of changes and variations. Most of you know the story of the Scotch rustic who was quizzed by an English tourist, who surprised him at his mid-day meal of brose. The tourist asked him what he had for breakfast and supper respectively, and on getting each time the laconic answer "brose," he burst out in amaze: "And do you never tire of brose!" Whereupon the still more astonished rustic rejoined "Wha wad tire o' their meat!" "Meat" to this happy youth was summed up in brose, and to go without was to go unfed.

Well, I am afraid the most Spartan *hausfrau* among us will scarcely attain to such an ideal of simplicity, but we might do well to have one staple dish, either in plane of, or along with porridge. For this purpose I know of nothing better than

Shredded Wheat Biscuits.

These have been referred to several times already in various savoury recipes, and, indeed, the ways in which they may be used are practically unlimited. For a

Standard Breakfast Dish,

especially in these days of "domestic" difficulty, they are exceedingly useful. For some years now we have bought them through our grocer by the case of 50 boxes—which, of course, brings them in much cheaper than buying these boxes singly—and use them week in, week out, for the family breakfast. Most people are familiar with the appearance of these, but any who have not yet sampled them should lose no time in doing so. Fortunately, they can now be had of all good grocers. When some of us began to use them first we had no end of bother sending away for them to special depots.

To prepare:—Have a flat tin or ashet large enough to hold the biscuits side by side. Spread the tin liberally with butter, lay in the biscuits, put more butter on the top of each, and toast till nicely crisp and brown in good oven, or under the gas grill. If the latter, turn to toast the under side. Be very careful not to burn. If toasted on an ashet serve on same dish. One can now have fire-proof ware which is not unsightly. There is a very artistic white fire-proof ware which is specially suitable for using in this way, so that besides the saving of trouble, one can have the food hot and crisp from the oven—a rather difficult, or at least uncertain consummation if there is much shifting from one dish to another. These

"Shredders,"

as we familiarly dub them, are most toothsome served quite simply as above, but they may be acceptably varied with sundry relishes. A very good way is to have a little gravy prepared by diluting half a teaspoonful "Marmite" or a teaspoonful "Carnos" in a half teacup *boiling* water. Pour a very little over each biscuit, and serve on very hot plates. Prepared thus they may serve as toast for scrambled eggs or any savoury mixture. For

Tomato "Shredders"

fry the necessary quantity of tomatoes, free from skin and seeds, in a little butter, with seasoning of grated onion, pepper, and salt. A little "Marmite" or "Carnos" is a great improvement.

<b<Mushrooms

may be used in the same way, and a mixture of mushrooms and tomatoes fried or baked and mixed together is especially good.

Green Onions

are also very good. Take 1/2 lb. green onions, trim away any tough or withered parts, and cut up the green in 1/2 inch lengths. Put these in a saucepan with boiling water to barely cover, a little salt, pinch sugar, and a little mint, sage, or parsley. Cook gently for half an hour, then add the white cut in rings, and stew till quite tender. Stir in 1/2 oz. butter, a little ketchup or extract, and serve on prepared S.W. Biscuits.

Craigie Toast

will commend itself to those who wish for a quickly made dish. Allow one egg and a small tomato to each person. Beat up the eggs and add the tomatoes minced, also seasoning—a few capers or a little gherkin finely chopped is very good—and a little milk, ketchup and water, or diluted extract—half a teacupful to 4 eggs. Melt a good piece of butter in saucepan, pour in the other ingredients, and mix over the fire till thoroughly hot. Cover, and allow to cook by the side of the fire for a few minutes, then serve piled up on crisp toasted S.W. Biscuits.

All the recipes I have given for using these biscuits are designed to have them dry and crisp. I think they are much nicer in that way, but those who like them soft or as a mush can have them so with even less trouble. Put a little milk, tomato juice, extract, sauce, &c., &c., in a soup plate. Dip in each biscuit lightly and drain, place on buttered tin or dish to warm through. For a

Bachelor's Mush

which might suitably take the place of porridge where the preparation of that is inconvenient, toast one or two Shredded Wheat Biscuits on a deep plate. Pour boiling milk over and serve with sugar or stewed fruit.

With stewed fruit, also, one might use

Triscuits

toasted or plain. These are flat filamented biscuits which can be used to advantage in many ways. They can be used in place of toast, and are very suitable to eat with porridge or any food which may be rather mushy alone.

One might go on for pages with suggestions for using these handy biscuits, but one has only to begin using them to find out innumerable ways of one's own. These are not always what *I* would suggest. One "unreformed" friend of mine who had begun to use them on my recommendation, told me she put them to fry every morning, after dipping in egg or batter, among the fat of the breakfast bacon!

Grain Granules.

This also is a very handy and sustaining breakfast dish, and needs little or no cooking. To make a hot mush put a few spoonfuls in a plate or saucer, and pour hot milk over. It may be eaten at once or allowed to remain in the oven for a few minutes. If to be eaten with cream or stewed fruit, crisp for a few minutes in the oven.

Nutgraino

is another excellent breakfast dish, composed of the whole wheat berry blended with nuts, and is most nourishing and digestible. It may be used as Grain Granules.

Wheatose

is a food which is recommended by eminent authorities on the food question. It is not so quickly prepared as the foregoing foods, but with a little forethought costs very little trouble. One teacupful should be soaked with rather less than twice that quantity of water for 10 hours, then it should be steamed in Queen pudding bowl, "Gourmet" boiler, &c., for 4 or 5 hours. It might thus be put on to soak in the morning, then put on to steam in the evening, or it might be put in covered jar in the oven all night. It can easily be warmed up in the morning, and when cold it will be quite firm, and may be cut in slices and fried. As a mush it should be eaten with dry toast or triscuits and stewed fruit.

COLD SAVOURIES.

"Reform" Mould.

(Mrs W., Dundee.)

Take 1 lb. yellow lentils, wash well, and boil with as little water as possible and any suitable seasoning, such as chopped onion, pinch herbs, salt, pepper, and a little butter; also about 2 tablespoonfuls of tapioca which has been soaked all night or longer. Cook very gently till the tapioca is quite clear, and turn into wetted or oiled mould. Turn out when quite firm and serve with any suitable garnish—cooked beetroot, &c.

NOTE.—This can be best cooked in double boiler, as it is very ready to catch the pan.

Vegetable Mould.

Cut finely about 6 ozs. each of turnip and carrot, and 3 ozs. shallots, and stew till just tender in stock or gravy to barely cover. Steaming is better, as the vegetables should not be broken down. Add some cooked cauliflower cut small, a cupful of cooked green peas or French beans, and 3 or 4 tomatoes sliced and cooked. Mix in 2 ozs. bread crumbs, and the same of cooked savoury rice, semolina, or tapioca, and cook a little longer. Press into a dish—an oval cake tin does very well. When cool turn out, see that it is neat, and brush all over with glaze. Garnish with slices of hard-boiled egg and

Tomato Aspic.

This jelly comes in useful in many ways. Take 1 tin tomatoes and rub through a sieve. Make up with clear stock or water to 1 pint—2 breakfastcupfuls. Have 1/6-oz. Agar-agar (Vegetable Gelatine) soaked for an hour in cold water, pour off the water, add to the tomato pulp, and put all in enamelled saucepan along with any additional flavouring required. Salt and white pepper will do nicely, but a blade of mace, some mixed herbs, and a few Jamaica peppercorns may be used. Add also the whites and shells of two eggs, unless you have a number of egg shells, in which case the whites may be dispensed with. Whisk steadily over the fire till it boils, then draw to the side and allow to simmer gently for 10 minutes. Pour twice through jelly-bag. The second time run half on to a flat ashet or some plates. Colour the rest with a little carmine and put to set also. When used as a garnish, stamp out in pretty shapes, and arrange with the red and amber alternating. For

Glaze

dissolve 2 tablespoonfuls of the clear tomato aspic in saucepan. Add 1/2 teaspoonful "Marmite," or 1 teaspoonful "Carnos" extract, mix thoroughly, and boil up. Allow to get nearly cool, but not beginning to set, and then brush over the mould with it.

Mock Calf's Foot Jelly.

Prepare according to directions given for tomato jelly, and just before straining add amount required of a good extract. One oz. "Marmite"—or 2 teaspoonfuls—or 1-1/2 ozs. "Carnos" to a pint of tomato jelly, would be a good proportion. Stir till dissolved. Strain and mould in the usual way.

It may of course be prepared without extract, by making a good strong stock. Vegetables may be used or not at discretion. The liquor strained from haricots, brown beans, or German lentils, with vegetable gelatine, in the proportion of 1/8-oz. to the pint, makes a delicious jelly. Care must be taken to see that none of the pulp gets through. Clarify and strain very carefully.

Legumes en Aspic.

Get an equal quantity of red, white, and green vegetables—say young carrots, tomatoes, turnips, cauliflower, green peas, French beans, &c. Have each cooked "to a turn" separately, and the carrots and turnips cut into neat shapes, cauliflower in tiny sprigs, &c. Arrange the vegetables as neatly as possible in a mould, and fill up with tomato jelly. When set, turn out and garnish with slices of fresh tomato and lemon.

It is not necessary to have a number of different vegetables for this dish. Any one or two of them will do quite well. The mould might be decorated with slices of beetroot or hard-boiled eggs.

Tomato and Egg Savoury.

Boil hard 4 eggs, cut in half, and remove yolks. Divide 4 good-sized, firm, ripe tomatoes in halves, and scoop out some of the pulp, leaving a nice case. Put the half whites inside the tomato shells and fill with the following mixture:—In a saucepan melt 2 ozs. butter, add tomato pulp, 1 oz. fine crumbs, the yolks rubbed through a sieve, a teaspoonful extract, salt, pepper, and a little lemon juice. Mix well and make quite hot. Fill in the little cups, piling it up cone-wise, and serve on a bed of aspic.

Raised Haricot Pie.

Prepare a raised pie case (see Pastry), put in a layer of cooked haricot or butter beans, a layer of sliced tomatoes, and one of hard-boiled eggs. Put on the lid, which should have a hole in the centre. Bake, and with a funnel fill in with dissolved savoury jelly. This is delicious to eat cold, and is very useful for pic-nics. The same ingredients may also be made into small pies or bridies.

POTTED SAVOURIES.

There is an unlimited variety of these to be had. Any of the savoury mixtures given in previous recipes for stews, sausages, &c., will do, but if to be kept for any length of time, it must be well seasoned, the different ingredients thoroughly blended or pounded together, and the mixture pressed into small jars or glasses with clarified butter or pure vegetable fat poured over. A little lemon juice and grated lemon rind will give a piquant relish to most of these potted "meats."

Haricot Paste.

This is very good, and is a handy way of using up cold haricots, butter beans, &c. Drain away any sauce, or add as much finely mashed potato or cold boiled rice as will absorb it. Add seasoning to taste—mace, made mustard, ketchup, "Extract," &c. Mix thoroughly and pass through a sieve to remove skins, stringy portions, &c. Some tomato is always an improvement, and if none has been cooked with the beans, put some in saucepan with a little butter and cook for 10 minutes. Add the haricots, &c., blend together over the fire, and pass through sieve while hot.

Lentil Paste

is made by using cooked lentils in place of the beans.

Tomato Paste.

Peel and cut small 1/2 lb. tomatoes. Put in saucepan with 1 oz. butter, a teaspoonful grated onion, and seasoning to taste — made mustard, celery salt, lemon juice, ketchup, "Extract," &c. Each or all of these are good. Stir over the fire till the tomato is nearly cooked, then add one egg, and stir round till all is smooth and thick. Add 2 tablespoonfuls bread crumbs or 1 of cold cooked rice, macaroni, &c., previously put through a sieve or masher. Remove to side of fire and stir in 2 ozs. grated cheese. Mix very thoroughly and pot.

Tomato Paste (2).

For immediate use the following is specially good. It may be used as a savoury, and makes a delicious filling for sandwiches. Take some firm, ripe tomatoes, free from skin and seeds, and cut up small. Allow 1 oz. grated cheese to every 4 ozs. tomato — some may prefer more cheese in proportion, but that is a fair average. Put in a strong basin with seasoning — made mustard or pepper, ketchup, a little "Marmite" or "Carnos," &c., and pound to a smooth paste with a wooden spoon. Pass through a sieve, and it is ready for use.

Brawn for Pic-Nic.

Take a small teacupful lentils, haricots, or butter peas, and rub through a sieve. Cook 2 ozs. flaked rice or semolina in a teacupful boiling stock for about 10 minutes, stirring all the while, and then

1/2 lb. or more of tomatoes sliced and cut small, dessertspoonful grated onion, some finely shred cooked carrot or beetroot, and seasoning. Add the lentils to this and mix thoroughly. Cook for a minute or so, remove from the fire, and mix in 2 finely chopped hard-boiled eggs. Press into a glass dish. It may be covered with glaze when turned out, or decorated with aspic jelly.

Tomatoes and Mushrooms,

gently baked or steamed together, with butter and seasoning, are also very good as a cold savoury for sandwiches; &c. If rather moist add a little cooked rice, mashed potato, or fine crumbs. Pound together, pass through a sieve if wished very smooth, and pot as above.

Sandwiches.

A good filling for sandwiches is to be found in any of the "potted meats" given in the foregoing section. Amongst others are

Egg Sandwiches.

These are usually made with finely chopped hard-boiled eggs. The latter alone may be used, or a little relish of some sort may be added—ketchup, tomato pulp, or chutney. Mix all to a smooth paste before using, and spread very evenly.

Egg Sandwiches (2).

Another very good way is to beat up the eggs a little, add seasoning, &c., put a bit of butter in saucepan, pour in the eggs, and cook gently till set. Stir all the time. Use when cold.

Water-Cress, Mustard-and-Cress,

and all salad vegetables are suitable for sandwiches. Most people will prefer them simply with bread and butter, so that the individual flavour may be appreciated. If any, such as lettuce or endive, are considered rather insipid, a little relish may be added as above. A tasty and novel flavour is obtained by spreading a very little Marmite Extract on the bread and butter before adding the filling proper.

Tomato Cheese Sandwiches

are among the best. The filling may be either the Tomato Paste given under Potted Savouries, or the mixture given for Scotch Woodcock or Mock Crab.

VEGETABLES.

It may seem rather supererogatory to speak of "Vegetables" distinctively, for the "unregenerate" will be inclined to declare that we have been discussing nothing else all the while. But for the benefit of such as are like the advertised domestic "willing to learn," I would say that vegetarians as a rule use fresh vegetables practically in the same way as meat eaters do, to supplement more substantial viands. No one—to my knowledge at least—ever dines off the proverbial cabbage or turnip—perhaps it would be better if they did now and then—but, that by the way. But there are vegetables *and* vegetables. No one who has gone in for the most elementary food reform will tolerate the sodden, soap-like potatoes, or the flabby, insipid, brown papery-looking stuff, called by courtesy cabbage, which so often does duty as companion to beef, mutton, or pork. Perhaps, though, the savoury cow or pig throws a halo over all the defects of its surroundings. Be that as it may, there is need for improvement in many ways, and by this I do not mean more elaboration in dressing or serving, for this is not seldom used to disguise shortcomings which otherwise could not escape notice. But disguising defects does not remove them, and we should do well to safeguard ourselves by having our food cooked as simply and naturally as possible.

The homeliest vegetables, too, if sound, ripe, and wholesome, are infinitely to be preferred to the rare expensive sorts forced out of season or gathered barely ripe and conveyed long distances to whet jaded palates. Well, to begin with that vegetable we are supposed to live on,

Cabbage.

This may either be a choice delicacy or an unmitigated abomination. It should be fresh, green, crisp and tender, and as newly pulled as possible. Those who have gardens should leave it growing till half-an-hour before cooking. When it must be kept for some time, see that it is in a shady, cool place, and an hour or two before using; remove any tough or withered leaves, split up the stalk well into the heart, if to be used whole, and lay in a large basin of cold water. Add a handful of salt and two tablespoonfuls vinegar to each gallon of water. Although freshly pulled all leafy vegetables should be soaked in this way to remove any caterpillars, slugs, &c., for even eaters of pig and ox have a curious objection to animal food on a small scale. To cook, have ready a good-sized saucepan with fast-boiling water containing a little salt, and if the cabbage is at all old or tough, a bit of washing soda the size of a hazel nut, to each quart of water. Drain very thoroughly from the water in which soaking, and plunge into the fast-boiling water. It is most important that the water should not go off the boil as then the juices would be drawn out and wasted. Boil steadily with the lid off for 10 to 20 minutes according to age, then lift into drainer on top of the boiling water and cook till tender in the steam. Serve on hot vegetable dish with some bits of butter on the top. It should be perfectly tender, yet crisp and of a vivid green. If at all brown, or dull, or flabby-looking, there is something wrong, either with the vegetable itself or the cooking. And I am not to give directions for "doctoring" anything that is either unwholesome or spoiled. A paragraph has been going the round of certain papers lately, giving directions for disguising the flavour of tainted meat, which "few cooks know how to treat so as to render perfectly nice"! It is to be wrapped in vinegar cloths, &c. — "boil up, and use it." I should say doctor it as you please, but then — throw it away! If anything, no matter what, goes bad — milk, soup, vegetables — throw it out without hesitation. It is a pity to waste things — and this ought to be prevented by good management — but surely it is much greater waste to use tainted food. Better miss a meal, if need be, than make a refuse bin of our bodies. All this may seem a digression, but I am so thoroughly convinced that a large proportion of the "ills that flesh is heir to" — and we accept the inheritance with a resignation "worthy of a better cause" — is due to unsound or improperly prepared food, that I make no apology. Many people have told me that they daren't touch certain vegetables, and when I have seen these as served by them have cordially agreed with them. The most common error, especially with green vegetables, like

Cabbage, Savoys, Brussels Sprouts, Greens, &c.,

which all require much the same treatment, is over-cooking. There seems to be a popular notion, somehow, regarding vegetables, that the more you cook them the better they are, and after all the substance and flavour has been boiled out of them, people wonder how anyone can relish such stuff! Each vegetable should get just the bare amount of cooking necessary, and no more. If they have to wait for some time before serving, stand over boiling water as directed above. Most vegetables may be cooked entirely by

Steaming.

This conserves all their own juices which contain the various valuable natural salts, alkalies, &c., so necessary to health, and which we so vainly try to make up by the addition of crude minerals.

Carrots, Turnips, Potatoes,

and all root vegetables and tubers, are best cooked by steaming. Steamers with perforated bottoms to fit the various sizes of saucepan are now to be had from any ironmonger. A very good way to cook carrots, turnips, and parsnips, is to make up a good white sauce, put in Queen pudding-bowl or some other such dish, lay in the carrots, parsnips, &c. Cover and steam till cooked. If rather old, they may first be par-boiled. This should be done before the skin is removed.

Beetroot

should always be steamed by preference, but quite as much care must be taken not to break any of the fibres, or it will "bleed" as in boiling. When tender, which will take from two to four hours, pare and cut in slices. It may either be dressed with vinegar, lemon juice, &c., to serve cold, or fried and served with white or tomato sauce as a hot vegetable.

Green Peas

may also be steamed in a jar or basin like stewed fruit. A very little water and a little lemon juice should be added. If to be boiled, have a small saucepan with fast boiling water to barely cover, a little sugar, salt, lemon juice, and sprig of mint. Boil fast till tender. Drain and serve with butter only.

French Beans

may be cooked in same way. Remove stalks and "strings" and cut across diamondwise.

Broad Beans, Kidney Beans, &c.,

usually require to have the tough white sloughs removed. To facilitate this, pour boiling water over, when they may be slipped off quite easily. Cook same as green peas.

To Re-heat Peas, &c.,

Put a little butter in saucepan, a finely minced shallot or spoonful grated onion, and some tomato free from skin and seeds. Simmer till cooked, lay in the vegetables to be warmed up. Make thoroughly hot and serve.

Cauliflower.

Trim and lay in cold salt water for some time, then boil or steam till just done. Trim off all the green leaves—it is best not to do this before cooking, as it is not so ready to break—lay in vegetable dish, and pour white sauce over.

Cauliflower au Gratin.

Prepare exactly as above. Coat with the sauce, sprinkle all over with bread crumbs or grated cheese, or a mixture of both, put some butter in little bits over it, and bake a light-brown in moderate oven.

Artichokes.

These may be cooked same as cauliflower, but require longer time. Cut the stalk off quite bare, and trim the leaves with scissors where necessary. By way of variety the centre part may be removed and the cavity filled with forcemeat or sausage filling. Serve with white sauce.

Jerusalem Artichokes.

Wash well, pare neatly, and lay in cold water and vinegar to cover. Have ready some boiling water with a little salt and some milk. Boil gently till tender—15 to 20 minutes. Drain, and serve with white sauce.

Fried Artichokes.

Parboil lightly, dry, dip in beaten egg, then toss in bread crumbs or a mixture of crumbs and grated cheese. Fry in smoking hot fat, and serve very hot on a napkin.

Fried Celery.

Prepare exactly as above. The pieces should be about 5 or 6 inches long.
Pile up crosswise in serving.

Stewed Celery.

Wash and trim the celery into short lengths and allow to soak in vinegar and water for an hour or so before cooking. Drain, and parboil in water containing a little salt and lemon juice or vinegar for 10 minutes. Drain again, and stew for another 10 or 15 minutes in some good white stock. Do not throw away the water in which celery, cauliflower, peas, &c., are boiled. It can be added to the stock-pot. Meantime toast a slice of bread, dip it in this celery water, and lay on ashet cut in triangles. Lay the celery on this when cooked, make the stock in saucepan into a good sauce with flour and butter, and pour over.

Seakale

is rather scarce and expensive as a rule, but it is well to know how to cook it when occasion offers. It is a choice delicacy for an invalid or convalescent. Soak in salted cold water for a time, trim neatly and cook till tender—about half-an-hour in fast boiling water containing a little salt and lemon juice. Drain, and serve on toast with white sauce over.

Asparagus.

Wash well in cold water and scrape the stalks white. Tie in small bundles and stand in fast boiling salted water till the stalks are tender—about twenty minutes. Drain, and serve like celery.

Salsify,

or vegetable oyster, is another vegetable which would find great favour were it not so scarce and dear. Scrape the roots and throw into cold water. Cut in 2-inch pieces and simmer gently for an hour or till tender in stock with a slice of lemon, or in milk and water. Lift out the salsify and place on toast. Thicken the liquor with butter and flour and pour over.

All vegetables which are served with white sauce or melted butter can be acceptably served

Au Gratin,

and a dish of carrots, turnips, and the like served in this way is quite a delicacy. Young tender vegetables are of course always to be preferred, but even when rather old are better this way than any other. Cook till quite tender, but not in the least broken. Lay in a pie dish, cover with sauce, coat thickly with crumbs or cheese and crumbs. Dot over with butter, and bake a light brown.

Spinach.

Soak in cold water and rinse very well to remove all grit, &c. Trim away stalks and tough fibre at the back of the leaf. Shake the water well off, and put in dry saucepan with lid on, to cook for about 10 minutes. Drain, chop finely, and return to saucepan with some butter, salt and pepper, to get quite hot. Dish neatly in a flat, round, or oval shape, with poached eggs on top, and croutons of toast or fried bread round.

Cauliflower — Dutch Way.

(Mr VAN TROMP.)

Boil cauliflower in usual way, drain, and put in vegetable dish. Coat with this sauce: — Make a cream with 2 spoonfuls potato flour, add a little sugar, and stir over fire till it thickens.

SALADS.

"Cucumbers, — Peel the cucumber, slice it, pepper it, put vinegar to it, then throw it out of the window." – *Dr Abernethy*.

One does not need to be a vegetarian to appreciate salads, and many who find cooked vegetables difficult of digestion, will find that they can take them, with impunity, raw, but it is inadvisable to take raw and cooked fruit or vegetables at the same meal.

Raw Cabbage,

for example, digests in little over an hour, while cooked it takes 3 to 4-1/2 hours. Needless to say, only young, tender, freshly pulled cabbage can be used in this way. Shred finely, removing all stalks and stringy pieces, and cover with the usual salad dressing. This may now be had ready for use in the shape of

Florence Cream,

but if wanted to be made at home, take equal quantities of finest salad oil and either lemon juice or vinegar and mix together gradually by a few drops at a time. A little cream or yolk of egg beat up is an improvement, and ketchup, made mustard, &c., may be added to taste. The dressing may be prepared beforehand, but should be put on just before sending to table.

Cold Slaw

is a favourite American salad. Shred the cabbage as above and sprinkle liberally with salt. Allow to remain for at least 24 hours, turning occasionally. Drain and use with lemon juice or salad dressing.

Tomato Salad.

Shred down a crisp, tender lettuce. Put in salad bowl. Scald and pare some firm, ripe tomatoes. Slice and cut up—not too small. Mix with lettuce. Pour over a simple dressing. Some slices of hard-boiled egg may be used as a garnish, or the white may be chopped up and the yolk grated over at the last. Tomato aspic is also a tasteful addition. Chop up and put lightly over. This salad or plain lettuce may be varied by adding almost any tender young vegetable, shred fine. Scraped radish, young carrots, turnips, cauliflower, green peas, very finely shred shallot or white of spring onion, chives, cress, &c., are all good, and may be used according to taste and convenience. A good

Winter Salad

can be made with celery, endive, &c., and of course with cold cooked vegetables. These latter should be cooked separately, and mixed tastefully together with an eye to colour and appearance. Raw and cooked vegetables should never be mixed in the same salad, or indeed eaten at the same meal.

SAUCES.

"Hunger is the best Sauce."

"England" has been slightingly defined by a French gourmand as a country of fifty religions and only one sauce! If this be true of those who have all the resources of the animal kingdom at their disposal, what can be the plight of those from whom these are shut out. This "one sauce" was, I believe, melted butter, or as it is more generally now called

White Sauce,

and it is not every one who can make even that plain sauce as it should be. The thin, watery mixture, or grey "stodgy" mass which is sometimes served with cauliflower or parsnips, even where the other viands are fairly well cooked and served, is certainly enough to condemn "vegetables." Yet, how simple it is if done the right way. In a small saucepan—preferably earthenware or enamel, for it must be spotlessly clean and smooth—melt 1 oz. butter, and into that stir 1 oz. flour. When quite smooth add by degrees a teacupful milk. Stir till it thickens, and allow to cook for a minute or two longer. It must be done over a very gentle heat—the side of the range, or gas stove turned low. If wanted more creamy, use more butter in proportion to the flour. Salt and pepper to taste. To make

Parsley Sauce,

add a spoonful of finely chopped and scalded parsley to this just as it comes a boil; and for

Caper Sauce,

add some finely chopped capers or fresh nasturtium pods in same way.

Tarragon Sauce.

Add 20 to 30 drops Tarragon vinegar to prepared white sauce. Stir well.

Dutch Sauce.

To a creamy white sauce made with 2 ozs. butter to 1 oz. flour, add one, two, or three yolks of eggs according to richness desired. Beat up a little, add a very little cold milk to prevent curdling. Stir into sauce when off the fire. Allow to come just to boiling point again—this should be done in double saucepan or boiler—and add a little lemon juice.

Dutch Sauce (2).

Take the yolks of 2 eggs, beat lightly, and add to them a teaspoonful cold water. Whisk in a saucepan, add a tablespoonful lemon juice, same of cream, and a little pepper and salt. Stir over slow heat till it thickens.

Egg Sauce.

Prepare white sauce as above, and when ready add one or two hard-boiled eggs, very finely minced. The sauce may be made with white stock instead of milk. A pinch cayenne and other seasoning may be added.

Celery Sauce.

Make a sauce with the water or stock in which a head of celery has been boiled. Pulp part of the finest of celery through a sieve and add.

Horse Radish Sauce.

To quantity required of white sauce, add one or two tablespoonfuls finely scraped horse radish, and the juice of a lemon or a little vinegar.

Mustard Sauce.

Add teaspoonful or more made mustard to each 1/4 pint white sauce.

Onion Sauce.

Boil 1/2 lb. or 3/4 lb. Spanish onions in milk and water till tender. Drain and make sauce with the liquor. Rub the onion through sieve and add.

Brown Sauce.

With brown stock or gravy, make a sauce in same way as white sauce. If browned flour is used the colour will be better. Add also a little Carnos or Marmite.

Hasty Brown Sauce

can also be made by using water, in which a teaspoonful Carnos or 1/2 teaspoonful Marmite to the teacupful has been dissolved, instead of the brown stock. Some mushroom ketchup is a good addition.

Sauce Piquante.

Stew some shallots in butter till quite cooked. Stir in a dessert spoonful flour and allow to brown. Add juice of a lemon and seasoning of cayenne, clove, &c., or a spoonful Worcester or other sauce, also 2 teacupfuls diluted extract or ketchup and water. Boil gently for 10 to 15 minutes, then strain.

Walnut Gravy.

This excellent sauce will be new to many, and some who, like the immortal "Mrs Todgers," are at their wit's end to provide the amount of gravy demanded, "which a whole animal, not to speak of a j'int, wouldn't do," may be glad to give it a trial. Take 2 ozs. grated walnuts. These should be run through a nut mill. Make 1 oz. butter hot in saucepan, add the walnuts and stir till very brown, but be careful not to burn. Add a tomato peeled and chopped, or a little of the juice from tinned tomatoes, a teaspoonful grated onion, and a very little flour. Mix well over the fire, and add slowly a breakfastcup brown stock or prepared Extract. Simmer gently for about 20 minutes. It may be strained or not, as preferred.

Tomato Sauce.

Peel and chop up 1/2 lb. tomatoes, or take a cupful tomato pulp. In a saucepan melt 1 oz. butter and add a little grated onion and the tomatoes. Simmer till cooked. Stir in a little flour or cornflour, and when that is cooked rub through a sieve. A little ketchup or lemon juice may be added to taste.

Mayonnaise Sauce.

Put the yolk of an egg in a basin and mix in a teaspoonful mustard and 3 or 4 tablespoonfuls salad oil, by a few drops at a time, beating all the while with a fork. Add the juice of a lemon, a little Tarragon vinegar and castor sugar, pinch cayenne, and if liked, the white of egg beat stiff, or a little cream at the last.

Mint Sauce.

Melt 1 tablespoonful castor sugar in a gill boiling water. When cold add same quantity vinegar, then 3 or 4 tablespoons freshly pulled mint, chopped small.

Curry Sauce.

Add 2 teaspoonfuls curry powder or paste and a little chutney to 1/2 pint
Brown Sauce or Piquant Sauce.

Bread Sauce.

Put a teacupful fine crumbs in a basin, add a tablespoonful grated onion, and pour over 2 cupfuls white stock or milk and water. Let stand for a little with plate over, then cook gently till quite smooth. Add seasoning of white pepper, ketchup, mace, &c., and if wished very smooth add a yolk of egg or a little cream, and rub through a coarse sieve.

Sweet White Sauce.

To 1/2 pint melted butter add 2 ozs. sugar and a little of any flavouring preferred. A yolk of egg beat up is an improvement.

Cocoanut Sauce.

To above sweet white sauce add when cooking, 2 ozs. cocoanut cream. Stir till dissolved. A little dessicated cocoanut will do, but the cream is much handier and nicer, as one has the rich cocoanut flavour without the tough fibre.

Almond Sauce.

1/4 lb. fresh butter or 3 ozs. almond butter, 2 ozs. sifted sugar, 1 oz. almond meal, or same of almonds blanched and chopped, 2 tablespoons water, 2 teaspoonfuls lemon juice.

Beat butter and sugar to a cream. (It should be quite light and frothy.) Add water and lemon juice by a drop or two at a time while beating. It should look like clotted cream. Sprinkle the almonds over. Excellent with pudding or stewed fruit.

Lemon Sauce.

Make a teaspoonful cornflour smooth in saucepan with a little cold water.
Add a gill of boiling water, juice of a lemon, and 2 ozs. sugar. Let boil
a minute or two. If flavour of rind is liked, grate that in. Add a little Carmine to colour.

Apple Sauce.

Pare, core and mince 4 to 6 apples. Stew in jar with moist sugar and a few cloves or bit of lemon rind. Remove the latter before sending to table.

* * * * *

CARNOS THE VEGETARIAN FOOD AND MEAT SUBSTITUTE,

Is the Best Article of its kind upon the market, being an appetising wholesome extract entirely soluble and free from fat. Send 4d. in stamps for 1-oz. Sample and full particulars to

CARNOS CO., Great Grimsby, Lincs.

N.B. – No chemicals used in the manufacture.

* * * * *

DAINTY COOKING!
Royal Pudding Mould
Pure Earthenware.

Prices—1-, 1/6, 2/-, 2/6

Saucepan Brush
Cleans a saucepan in a few seconds. Price 6d.

Pie Cup
Price 4d. each.

NO CLOTH. NO STRING.

Opened and Closed instantly.

Water kept out; Goodness kept in.

Gourmet Boiler

Cooks Porridge, Meat, Beef Tea, Jellies, Fruit, &c.

No Stirring; No Burning; No Waste. Prices—9d., 1-, 1/3, 1/6, 1/8, 2/-, 2/3, 2/6, and upwards.

Egg Beater

For frothing Eggs and Foaming Cream Prices— 9d., 1/-, 1/6, 2/-

Queen's Pudding Boiler

Prices—9d., 1/-, 1/6, 2/- 2/6, 3/-

Pudding Spoon

Handy to use; does its work well. Price 6d.

Stands inside any Saucepan

Egg Separator
Instantly separates the white from the yolk. Price 3d. each.

Complete List Free on application to
GOURMET & CO., Mount Pleasant, London, W.C.

* * * * *

THE "ARTOX" FLAVOUR

HAVE YOU HEARD OF IT?

It is that delicious, sweet, nutty flavour which you long for but seldom find. It is only to be found in

"ARTOX"

Wholemeal, which is made from the very finest wheat obtainable, carefully selected and blended, and ground by millstones in the good old fashioned way.

"ARTOX"

contains the whole of the wheat, so treated that the sharp, irritating particles of the bran, so prevalent in the ordinary meal, are rendered harmless and capable of digestion by the weakest stomach.

"ARTOX"

by a patent process is ground to such a marvellous degree of fineness that it can be used for all the purposes for which white flour is used. Therefore make all your Bread, Puddings, Cakes, Pies, and Pastry with "ARTOX." They will be much nicer, besides being more nourishing and satisfying, because "ARTOX" is a perfect natural food.

We have a dainty booklet—"Grains of Common Sense"—we should like to send you, crammed with novel and delicious recipes. It will be sent free on application.

"ARTOX" is sold by all the leading Grocers and Health Food Stores in 3 lb., 7 lb., and 14 lb. sealed linen bags, or 28 lbs. will be sent direct on receipt of P.O. for 4/6.

Send Post Card for Name and Address of Nearest Agent to

APPLEYARDS, Ld., ROTHERHAM.

* * * * *

BREAD.

One of the chief difficulties experienced by those trying to compass a complete scheme of hygienic dietary, is to get a pure, wholesome, easily digested, and, at the same time, palatable bread. We have long since exploded the idea that *whiteness* is a test of superiority, for we know that this is attained by excluding the most wholesome and nutritious part of the wheat and by the use of chemicals. Even when we use brown bread, we are by no means sure of having a wholemeal loaf, for it is as often as not merely the ordinary flour with some bran mixed in. And bran is only one part — by no means the most important — of that in which the meal is lacking. We want to get as much as possible of the real "*germ*," the essential part of the grain, but I am informed by experts, that the process of drying and preparing this germ meal is so much more troublesome, and, in consequence, expensive, that the easier and cheaper method is that generally adopted. But, if we want a really good thing we must be willing to pay for it, and by creating a demand for the superior article make it worth while to manufacture it, and it were poor economy to save on the bread bill at the expense of health. It is well to know exactly what constitutes a really wholesome bread, for bakers and purveyors everywhere are ready to meet their customers' wishes. But if people are ignorant or unreasonable enough to demand a light-coloured, puffy loaf, when a pure whole-wheat loaf is rather dark and solid looking, they must be prepared to find that they are served with what pleases their taste, and to take the risks. Some may like to try baking their bread at home, and it may interest them to know that it is possible to make very good wheaten bread without any raising ingredients whatever, simply with wheatmeal and water, aerating it by beating air into it. This is best managed by the home baker in the form of

Wheatmeal Gems.

There are sets of thick iron gem pans to be had, which are very good for this purpose, but one can manage quite well with oven-plates made of sheet-iron or black steel.

Into a large basin put 2 cupfuls of the coldest water procurable. Aerate by pouring from one vessel to another several times, or by whipping up with a spoon or spatula. Take 4 cupfuls whole meal, and pass several times through a sieve. Sprinkle the meal into the water a little at a time, whipping vigorously all the while till about three-fourths are worked in, and continue whisking from 20 to 30 minutes till the mixture is full of air bubbles. Sprinkle in the rest of the wheatmeal and mix thoroughly. Meanwhile, see that the oven is very hot, as a strong steady heat is necessary. Make the gem pans or oven-plates also very hot and grease lightly. Half fill the pans and put at once in oven, so that the moist air may be as quickly as possible converted into steam, and thus puff up the bread. If oven-plates are used, put dessertspoonfuls some distance apart on these and put in oven. If the oven is hot enough, a crust will at once form, and the steam trying to force its way out will send them up like puff balls. Moderate the heat, if possible after 10 or 15 minutes, and allow to bake for about 30 minutes longer. It is very easy to regulate the heat if a gas stove is used; if a range, put on some small coal. When baked turn out on a sieve, and when quite cold split open and toast on the inside.

Another excellent kind of bread, which can be managed quite easily with a little trouble and practice, is raised with eggs. It is generally known as

Wallace Egg Bread,

and as I have the recipe direct from Mrs C. Leigh Hunt Wallace, the inventor of this kind of bread, I am able to pass it on at first hand.

Ten ounces wheatmeal, 1 large egg (weighing 2 ozs.), 1 gill milk and 1 gill water, the whole to be made into a batter, the white of egg being beaten separately to a stiff froth and incorporated with the batter very thoroughly but very quickly; the whole to be baked in 1 lb. cake or loaf tin, the tin being very hot and thoroughly oiled or buttered before the batter is turned into it. Put for 50 minutes in a very hot part of the oven (350 degrees to 380 degrees fahr.) and keep in another 50 minutes to soak. I can vouch for the excellence of this bread, and may say that I have managed it with very little difficulty. I use a gas oven and loaf pans made of black steel, as these take and retain the heat much better than tins. If any amateur, however, is doubtful as to how this loaf should be, she cannot do better than send for a sample loaf or two to the Wallace Bakery, 465 Battersea Park Road, London, S.W. There is also a depot in Edinburgh — Messrs Richards & Co., 7 Dundas Street, where these can be got. By comparing one's own achievements with these, one will be the better able to attain the desired result. In case any may think this egg bread sounds expensive, I may say that it is exceedingly economical to use; a small loaf going much farther than a large one of the ordinary puffed-up kind.

PASTRY.

"'Meat for Repentance' — Pork pies for supper — or otherwise!"

Short Crust.

Take 1/2 lb. flour, mix with it 1/2 teaspoonful baking powder and put two or three times through a sieve. Rub in 4 ozs. butter. If vegetable butter is used, 3 ozs. will do, as it contains much less water. Beat up an egg. Add a teaspoonful lemon juice to the flour, &c., nearly the whole of the egg, and mix into a very dry paste with cold water. The mixing is best done with a knife. Turn out on floured board and form into an oblong piece, still using a broad knife as much as possible. Roll out evenly a good deal larger than the dish to be covered, and cut off a piece all round, leaving it the exact size and shape. Wet the edges of the dish, put a band of paste on. Wet that again, and lay on the cover. Make the edges neat with a knife or pastry cutter. Brush over with egg and bake in very hot oven for thirty to forty minutes. If used for covering a fruit tart, dust over with sifted sugar before serving.

Rough Puff Paste.

Take same quantities as for short crust. Divide butter into pieces on floured board and flatten with the rolling-pin — a stoneware bottle, by the way, is much better than a wooden rolling-pin. Put the butter with the flour and mix as before with egg, lemon juice and water. Turn out on floured board, make into a neat, oblong shape, beat down with rolling-pin and roll out very evenly to about 1/8-inch thickness. Dust with flour and fold in three, turn half round so as to have open end in front of one, and roll out as before. Repeat this until it has got 4 turns, taking care to keep the edges as even as possible, and for the last time roll out a good deal larger than the dish. Put a band of paste on the dish, wet this and lay on the cover. Flute the edges neatly. Brush over with egg. Cut the trimmings of paste into leaves, &c., and decorate the pie, putting a rose in the centre. Brush these also with egg. Make one or two slits to let out the steam, and bake in hot oven. The oven should be made very hot *before* the pastry is put in, and then the heat should be moderated. This can of course be managed best with a gas oven.

This rough puff paste is very suitable for small sausage rolls. Roll out for last time quite square. Divide into nine equal squares, put a small quantity of sausage meat on centre, wet edges and press together. Brush over with egg and bake. Remember never to brush the edges with egg, as that would stick them together and prevent rising.

Rich Puff Paste

suitable for patties, vol-au-vent, &c., is made as above, but with 6 ozs. butter to 8 ozs. flour. For patties leave the paste at last rolling out 1/2 inch thick. Stamp out into rounds with lid or biscuit-cutter, about 2-1/2" or 3" diameter, and with a smaller cutter mark about half-way through the paste. Brush with egg and put on oven-plate. See that the oven is specially hot, and yet regulated so that the pastry will not scorch before thoroughly risen, as the oven door must not be opened for fifteen to twenty minutes after putting in. They should rise to three or four times the thickness of the paste. Allow to bake some time longer, remove from oven, and with a sharp-pointed knife remove the centre lid. Fill in with the mushrooms, tomatoes, &c., replace top, and make very hot again before using.

Vol-au-Vent

is done exactly in same way, only all in one. Cut out the whole of the paste round, oval or square, and with a sharp-pointed knife mark half-way through all round about an inch from the edge. Bake as for patties, but the larger piece of pastry will require longer to bake through and through. Remove lid carefully, put in filling and replace lid.

Raised Pie Crust.

This paste is most wholesome and economical. For a good-sized pie take 3/4lb. flour and 3 ozs. butter or Nut Butter. Put the flour in a basin. Bring the butter to boiling point with a teacupful water. Pour in among the flour, stirring all the time till thoroughly mixed, then knead well. When nearly cold take off about a third and make the rest into a ball, flatten and work up by hand till the case is about 2-1/2 inches high, and slightly narrower at the top—Melton-Mowbray shape. Slip on to greased oven-plate, and when quite firm, fill rather more than half-full with haricots, tomatoes, &c. Roll out the bit of paste remaining, cut out lid, wet the edges of it and the pie-case and pinch together. Brush all over with egg. Ornament with the trimmings, brush again and bake in good steady oven for at least three-quarters of an hour. When ready, pour in some more gravy, or if to be used cold, some dissolved savoury jelly.

Should there be difficulty at first in raising this entirely by hand, it might be moulded round a jar or round tin. Another way is to use a tart ring, but a very simple and handy way, which finds favour especially with children, is to make bridies. Divide the paste into ten or twelve pieces. Roll out a nice oval, put some savoury mixture on one half, wet edges with egg or water, press together and pinch into neat flutes, brush over with egg and bake.

Suet Paste.

Allow 3 ozs. vegetable suet to 8 ozs. flour. Chop the suet or run through nut-mill. Add to flour along with salt and pepper, and if liked, a little grated onion and chopped parsley. Make into a firm paste with water, which may have a little ketchup or "Extract" diluted in it.

This is, of course, for savoury pies, &c. If for sweet dishes—roly-poly, apple dumpling, &c.—omit all seasonings and add sugar and any flavouring preferred, such as clove, ginger, or cinnamon.

CAKES, SCONES, &c.

Only a few cakes, &c., are given here, as there are a number of excellent ones among the contributed recipes in last section, under heading of Bazaar contributions, and, besides, there is nothing about them peculiar to food reformers. Those who are studying wholesomeness and digestibility, however, will avoid as far as possible the use of chemicals for raising, and fats of doubtful purity such as hog's lard. The injurious character of carbonate of soda, tartaric acid, &c., if used at all to excess, is now fully recognised, and those whose health is not quite normal should avoid them entirely. When such cannot be dispensed with, use very sparingly and in the exact quantities and proportions of acid and alkali, which will neutralise each other by converting into a gas which passes off in baking, if the oven, &c., is all right. But the latter point is rather a big and very essential "if," and many cooks try to make up for deficiencies in mixing and firing, by putting in an extra allowance of baking powder. There is considerable diversity of opinion still as to the exact nature and place of these chemicals in the economy of the body, and where "doctors differ" the amateur cook or hygienist dare hardly dogmatise, but all are agreed that the slightest excess is hurtful. Cakes, scones, pastry and the like, should depend rather for lightness upon cool, deft handling, and skilful management of the various details which contribute to successful baking.

A fine essential is to have good, reliable flour. See that it is perfectly dry, and pass several times through a fine sieve to aerate and loosen it. Try to bake in a cool, airy place, and be provided with all the necessary tools for accomplishing the work in expert and expeditious fashion, for the success of many things depends upon the celerity with which the process is performed. Have the oven at just the right heat, at the right time. A cake which would otherwise be excellent may be heavy or tough by having to wait till the oven cools down or heats up to the proper temperature. With a gas oven, one can regulate at will, and a safe general rule is to have the oven thoroughly hot *before* the cakes are put in, and then to moderate the heat very considerably. With a coal fire, if the oven is too hot, put on a quantity of small coal.

Artox Gingerbread.

One and a half pounds Artox wholemeal, 10 oz. golden syrup, 9 oz. butter, 4 oz. sugar, 1/2 oz. carbonate of soda, 1/2 oz. ginger, 2 eggs, little milk. Cream together the butter and sugar, add the eggs, well beaten, and the syrup, stir until dissolved. Add the Artox wholemeal with the soda and ginger previously sifted in, and a little milk if necessary, to make a stiff batter. Put into greased tins, and bake in a moderate oven.

Artox Seed Cake.

Beat 10 ozs. of fresh butter to a cream, add 6 ozs. sugar and beat into the butter. Separate yokes and whites of 4 eggs and beat each mass separately. Then mix well with the butter and sugar, adding the yokes first and the whites last. Add 1 teaspoonful carraway seeds and 10 ozs. Artox wholemeal. Mix thoroughly, put into butter papered tins and bake in a quick oven.

Artox Shortbread.

One and a quarter pounds Artox wholemeal, 10 ozs. butter, 4 ozs. sugar, 1 egg, 1/4 oz. baking powder. Rub the Artox wholemeal, sugar, and butter together, add the baking powder, and make into a stiff paste with the egg. Mould it into cakes, crimp the edges, and bake in a moderate oven.

French Layer Cake.

1/4 lb. butter or fine nut butter. Four eggs, 1/2 lb. flour, 6 ozs. fine sugar, 1/2 teaspoonful baking powder, 1/2 teaspoonful essence vanilla, 4 ozs. grated chocolate, 2 ozs. icing sugar.

Butter 3 sandwich tins. Dissolve 1 oz. chocolate in pan, with 1 tablespoonful milk, over the fire. Beat butter and sugar to a cream. Beat up eggs very light, laying aside one white for icing, and add. Sift flour and baking powder, and mix in, then flavouring. Put a third in one tin, another in pan with chocolate, and put a few drops carmine in that left in bowl. Put these into the different tins and place at once in hot oven. They should be ready in 10 minutes. Put remaining chocolate with the icing sugar in pan with a tablespoonful water. Boil a minute with constant stirring. Turn out cakes on a towel. Put half of chocolate mixture on one, put another on the top, then the rest of chocolate, and, last, the third cake. Coat with the following

Icing.

Beat up white of 1 egg till quite stiff. Mix in 6 ozs. icing sugar. Put on very smoothly with a broad knife dipped in water. Sprinkle over with grated cocoanut, or decorate with pink icing put through a forcing-bag.

Cocoanut Icing

might be used instead. Dissolve about one fourth of a square of cocoanut cream with a little boiling water. When cool mix thoroughly with half of the above icing.

Gingerbread.

1/2-lb. flour, 1 oz. good cocoanut butter, 1 oz. sugar, and same of syrup or treacle—if the latter use more sugar. Two ozs. stoned raisins or sultanas, 1 teaspoonful ground ginger, and same of mixed spice. Half teaspoonful baking powder. One egg.

Mix all the dry things. Rub in butter, then add syrup, fruit, and egg, and make into a thick batter with milk. Bake in moderate oven half-an-hour or longer. Very good, if made with half wheatmeal, or a proportion of oatmeal or rolled oats.

Jumbles.

1/2-lb. flour, 1/4 lb. butter, 2 ozs. sifted sugar, 1 egg. Pinch baking powder. Beat butter and sugar to a cream, add egg, well beaten, then flour, &c. Knead into a stiff paste, divide into 12 or more pieces, and roll out pipe-wise with the hands, about a foot long. Curl round, or form into letters, &c. Lay on floured oven plate. Brush with egg. Sprinkle with sugar, and bake 15 minutes in hot oven.

Orange Rock Cakes.

1/2-lb. flour, 2 ozs. sugar, 1 teaspoonful baking powder, 1 oz. butter or cocoanut cream butter,[Footnote: [see next footnote]] 1 egg, 1 orange.

Mix flour and sugar, rub in butter. Add yellow part of orange rind, grated, and juice, also the egg well beaten, to make stiff dough. Place a little apart on oven plate, with two forks, in rough pieces about the size of a walnut. Bake about 10 minutes in quick oven.

Dinner Rolls.

1/2 lb. flour, 1 oz. butter or nut butter, 1 egg, 1 teaspoonful baking powder, 1 gill milk, pinch salt. Rub the butter into flour, &c. Beat up egg, lay aside some for brushing, and mix in lightly with barely a gill of milk. Turn on to floured board, and roll out. Divide into a dozen or more pieces. Roll round with the hands. Shape into twists, knots, "figure eights," &c. Put on floured oven plate. Brush over with egg, and bake about seven minutes in very hot oven.

Afternoon Tea Scones.

1/2 lb. flour, 1 teaspoonful baking powder, 2 do. sugar, 1 do. butter or "Nutter." One egg. Mix dry things. Rub in butter, beat egg, and add with as much milk as make nice dough—about 1 gill. Roll out 1/4 in. thick. Stamp out with small cutter or lid. Brush over with egg. Bake 10 minutes.

Cocoanut Cream Scones

are made by adding 1 oz. cocoanut cream [Footnote: NOTE.—Cocoanut or almond cream butter may be used instead of ordinary butter in most recipes for cakes or sweets, and will give variety of flavour.], dissolved in a little of the milk, to the above. Let the "cream" be cool.

Artox Scones.

Two pounds Artox wholemeal, 1/2 lb. butter, 5 oz. sugar, 1/2 oz. cream of tartar, pinch carbonate of soda, 2 eggs, milk. Put the salt, soda, and cream of tartar, into the wholemeal, rub in the butter, stir in the eggs (well beaten), and enough milk to make a stiff paste. Divide the mixture into five, roll each piece out about the size of a cheese plate, divide twice across, place on a greased tin for 10 minutes, bake in a *hot* oven.

Artox Tea Biscuits.

One and a quarter pounds Artox wholemeal, 3 oz. butter, half teaspoonful baking powder, milk, pinch of salt. Put the wholemeal into a bowl, rub in the butter, add salt and baking powder, and enough milk to make a stiff paste. Roll out, cut into rounds, and bake in a hot oven.

German Biscuits.

1/2 lb. flour, 1/4 lb. butter, 1/4 lb. sugar, 1 egg, 1/2 teaspoonful ground cinnamon.

Rub in butter among flour and sugar. Add cinnamon. Make into a paste with the egg beaten up. Knead till smooth. Roll out thin and stamp into biscuits. Bake about 10 minutes on greased oven plate in moderate oven. Stick two together with a little jam, and ice with 4 ozs. icing sugar mixed with a little water. Dust with pink sugar.

PUDDINGS AND SWEETS.

As a number of favourite puddings and sweets also are given in the last section, it will not be necessary to give here more than a few supplementary ones, mostly introducing specialties which are not so well known as they deserve to be. Besides, all sweet dishes are vegetarian already for the most part, so that there is but little to "reform" about them. Of course, those who wish to have them absolutely pure will substitute vegetable suet or butter, and vegetable gelatine for beef suet and clarified (?) glue.

Almond Custard.

Two eggs, 1/2 pint milk, 2 ozs. Mapleton's almond meal, 1-1/2 ozs. sugar.

Beat eggs with sugar, add almond meal. Almonds blanched and pounded will do, but the meal is ready for use and costs less. Add the milk and a few drops of flavouring. Bake in slow oven till set, or stir till it thickens in jug or double boiler. This is specially good with stewed fruit. It may be made into

Custard Whip Sauce

by putting in saucepan and whisking over the fire till light and frothy. It must not boil.

Banana Custard.

Five or six bananas. Jam. Custard. Peel the bananas, which must be sound and ripe; split lengthways. Spread each half with jam—apricot is very good; put halves together. Lay in glass dish and pour almond custard, or cocoanut cream custard, over.

Cocoanut Cream Custard.

This is made same as almond custard, but using cocoanut cream instead of the almond meal. This cocoanut cream, which is put up in tablets, is exceedingly useful for almost every variety of pudding, icing for cakes, &c. It has only to be chopped down or melted, and serves the double purpose of giving flavour and substance.

Canary Pudding.

Four ozs. flour, 4 ozs. butter or 3 ozs. Table Nut Butter, 2 eggs, 3 ozs. sugar, 1 teaspoonful baking powder.

Melt butter in saucepan. Add the sugar and eggs beaten up, the flour and baking powder; lastly, 2 tablespoonfuls milk. Mix thoroughly. Butter well a plain mould, and put into it some jam or marmalade. Pour in pudding, cover with buttered paper, and steam for 2 hours.

Artox Queen Pudding.

2 oz. Artox bread crumbs, 2 oz. sugar, 1/2 pint milk, rind of half a lemon, 2 eggs, and a little raspberry jam. Boil the milk, pour over crumbs, and add yolks of the eggs, sugar and lemon rind. Bake in a greased pie-dish 20 minutes in a moderate oven, then spread over about 2 tablespoonfuls of hot raspberry jam. Beat up the whites of the eggs to a stiff froth and place over the jam, then put in oven for about three minutes to set.

Appel-Moes (Dutch Recipe).

Peel, core, and slice quantity of apples required. Stew or steam in covered jar with sugar and flavouring of cinnamon. Pulp through a sieve with whipped cream or as a sauce for steamed pudding.

Lemon Sponge.

Soak 1/8 oz. vegetable gelatine in a tumbler of water for an hour. Strain and put in saucepan with a tumbler fresh water and 5 ozs. loaf sugar. Stir till gelatine is dissolved. Add juice of 2 lemons, and strain through sieve. When cool add the whites of two eggs, and switch till quite light and spongy throughout—about three quarters of an hour. Put in mould, or when set pile up in rocky spoonfuls.

Lemon Cream Mould.

1 large lemon, 3 eggs, 6 ozs. sugar, 3/4 pint (3 teacupfuls) milk, 1/6 oz. vegetable gelatine.

Soak gelatine in cold water for at least an hour. Drain and put to come slowly to boil in the milk. Separate whites from yolks of eggs, and put the latter in large basin with the sugar and yellow part of lemon rind grated. Beat thoroughly and strain boiling milk over, stirring all the time. Return to saucepan, bring just to boil, and set aside to cool. Beat up whites of eggs very stiff and mix in lightly, adding the strained juice of lemon. Put in mould or glass dish, and set in cool place till quite firm.

Cobden Pudding.

Four ozs. grain granules, 2 ozs. sugar, 1 oz. cocoanut cream, 3 ozs. stoned raisins, 2 eggs, 3 gills milk.

Put grain granules, sugar, raisins, and cocoanut cream in large basin. Bring milk to boil and pour over. Cover and let stand till cool. Beat up yolks and add, and lastly the whites beaten stiff. Pour into buttered pudding-dish and bake in moderate oven for an hour.

JAMS AND JELLIES.

We have not space to go into these at any length. The following are one or two of my "very own," as the children say, which are voted a great success.

Apple Jam.

Take quantity required — say 7 lbs. — tart crisp apples. Wash well and dry. Pare and core, putting the trimmings in water to cover. Cut up the best of the apples into small pieces — not too thin — and set aside, also covered with cold water. Put on the trimmings to boil with some lemon rind and either a few sticks of cinnamon or some cloves. Simmer for an hour or longer, till all the goodness is drawn out, mashing freely with a wooden spoon. Turn into jelly-bag and allow to drain without pressure. Pour the water off the apples, measure that and the drained juice, and put into preserving pan. Measure the apple chips also, and add when the liquid boils. Allow 14 ozs. loaf sugar to each breakfast cupful, and boil till the apples are clear, but not broken down — about 20 minutes. Skim and pot as usual. If ginger flavouring is preferred, shave down about 6 ozs. preserved ginger, and add when the juice is put on to boil.

Marmalade Jelly.

Take 3 lbs. fruit — 6 bitter oranges, 3 sweet ones and 3 lemons. Remove the rinds and grate them small, or put through a mincer. Cut up the oranges, removing the seeds, which put in a tumbler of water. Cover the oranges, &c., with 17 tumblers cold water, and let stand for at least 24 hours. Put all in jelly-pan, including the water drained from the seeds, and let boil gently, for about 2 hours, mashing frequently with a wooden spoon. Let drain without pressure. Measure the juice, and to each pint allow 14 ozs. sugar, which add after the liquid boils. Boil fast for a few minutes, try if it will set.

Skim and pot. But the pulp must not be thrown out, for it makes an excellent, if rather homely,

Marmalade,

which comes in specially useful for steamed puddings, &c. Weigh the pulp, and allow equal weight of sugar. Boil gently, taking great care not to burn, till clear—20 to 30 minutes.

Green Gooseberry and Strawberry Jam.

This will be appreciated by those who find the ordinary strawberry jam rather sweet and heavy. Take equal quantities of gooseberries and strawberries—say 3 lbs. of each. Trim the gooseberries, which must be firm and freshly pulled, and wash well. Put on to boil with a teacupful water to each lb. of gooseberries, and boil for 10 minutes. Add the strawberries and the sugar lb. for lb., and boil for 20 minutes longer, or till it will "jell," as Meg would say.

Green Gooseberry Jam

is made with the gooseberries alone, prepared as above. A little grated lemon rind, &c., might be used for flavouring. Then if one is making

Green Gooseberry Jelly,

top and tail the fruit very carefully, removing every tough or discoloured one. Put on to boil, well covered with water. Add flavouring or not as preferred, and simmer gently for an hour or so. Drain without pressure. Allow 14 ozs. to pint of juice, and boil rapidly about 10 minutes. Allow 1 lb. sugar to each lb. of the pulp. Boil together for about 20 minutes, and this will give a very good, if rough and ready, jam.

Jelly without Boiling.

Everyone who can get good red or white currants should try making the jelly without boiling. I got the recipe from a friend many years ago, and can recommend it as a way in which the fresh flavour of the fruit is preserved to perfection. Wring the currants in usual way, and to each pint of juice allow 14 ozs. loaf sugar, which must be pure cane. I believe crystalised will do, but I have never tried it. Granulated or beet sugar will not do. Put juice and sugar in a strong basin and beat with the back of a wooden spoon till the sugar is quite dissolved, which will take about half-an-hour. Skim and pot. It should be quite firm by next day, and will keep for a year or longer — if it escapes consumption.

Bramble Jelly.

This is one of the finest preserves one can make — especially if we have gathered the fruit. The brambles should not be too ripe, but should have a good proportion of hard red ones. Wash well in cold water and put on with water to barely cover. Simmer gently for an hour or longer, bruising well with wooden spoon. Drain without pressure. Measure, and allow 14 ozs. sugar to pint, *i.e.*, breakfast cupful. Allow the juice to boil up well. Add the sugar, boil fast for a few minutes, skim and pot.

NOTE. — Only pure cane sugar should be used for preserves. Add the sugar when the preserve is boiling — nearly ready indeed. It only requires to be thoroughly dissolved and boiled through. This method goes far to prevent burning and loss of flavour.

* * * * *

The NEW VEGETABLE FOOD EXTRACT which possesses the same nutrient value as a well-prepared Meat Extract.

2 oz. pot, 7-1/2 d.; 4 oz. pot, 1/1-1/2; 8 oz. pot, 2/-; 16 oz. pot, 3/4.

The Ideal basis for high-class Vegetable Soups.

HORS CONCOURS

Universal Cookery and Food Exhibition 1907.

MARMITE

THREE GOLD MEDALS AWARDED

Cookery Schools and Teachers are invited to apply for Free Samples, Recipes, and full particulars to

THE MARMITE FOOD EXTRACT CO., Ltd., 59 Eastcheap, London, E.C.

* * * * *

WILL YOU TRY A CUP OF TEA

that, instead of injuring your nerves and toughening your food, is Absolutely Safe and Delightful? 2/2, 2/10, and 3/6 per lb.

THE UNIVERSAL DIGESTIVE TEA

is ordinary tea treated with oxygen, which neutralises the injurious tannin. Every pound of ordinary tea contains about two ounces of tannin. Tannin is a powerful astringent substance to tan skins into leather. The tannin in ordinary tea tans, or hardens, the lining of the digestive organs, also the food eaten. This prevents the healthful nourishment of the body, and undoubtedly eventuates in nervous disorders. On receipt of a postcard, The Universal Digestive Tea Co., Ltd., Colonial Warehouse, Kendal, will send a Sample of this tea, and name of nearest Agent, also a Descriptive Pamphlet compiled by Albert Broadbent, Author of "Science in the Daily Meal," &c.

AGENTS WANTED.

* * * * *

THE BEST SOUP THICKENER.

ROBINSON'S PATENT BARLEY

Also Best for Making BARLEY WATER, CUSTARD, BLANC MANGES, &c.

KEEN ROBINSON & CO., LTD., LONDON,

Makers of Robinson's Patent Groats for making Gruel.

* * * * *

BEVERAGES.

We have not space to go into the question of beverages at any length. A few good "drinks" are given under Invalid Dietary, and I would just say that the juice of a squeezed lemon, orange, or other fruit juice is much better than any effervescent or chemicalised beverage. There are, however, some excellent pure fruit-juices now on the market, among which one may mention

Pattinson's Fruit Syrups

and essences for various temperance drinks as being specially good. Many are proscribed on the score of health, &c., from the use of

Tea and Coffee,

but as these will remain first favourites for a long time to come, the first essential is to have them properly prepared, so that there is little if any ill effect. Where tea is most largely and constantly used, as in China and Japan, it is said to be quite innocuous. This may be partly owing to the more wholesome and rational way in which those people live, partly also to the finer quality of tea available, but very largely to the method of preparation. Various devices have been patented to save trouble in changing from one pot to another, but as most of these are rather complicated for daily use, we are glad to learn of a tea which can be prepared in the old comfortable handy way without any ill effects, and this boon seems to be furnished in the

Universal Digestive Tea,

prepared at the Colonial Warehouse, Kendal. By a process — which, by the way, is not kept secret — the tea is treated with oxygen in such a way that the hurtful tannin is neutralised, while none of the other properties are affected in any way. There is certainly no loss of flavour, and no difference that one can discern from the usual, but specially good tea — a fact which will appeal to ordinary tea-drinkers, of whom there are still a majority. For any further information regarding this tea, I would recommend readers to a little pamphlet compiled by Albert Broadbent, Esq., food specialist and lecturer, whose writings on the food question, &c., are well known. It is entitled "The cup that cheers." It explains the process of treatment, and gives medical and analytical testimony in its favour from various authorities of very high standing. The best proof is in the drinking, however, and one may have a sample pound or more carriage paid.

INVALID DIETARY.

The whole of the previous part of this book has been devoted to the contriving of the several meals usual in a work-a-day household and under ordinary circumstances. But exceptions will occur in the "best regulated families," and although much may be done to prevent illness by pure, nourishing, well-cooked food, one must be prepared for emergencies as they come.

Of course, most of our friends will be only too ready to pounce upon us when illness comes into the house, with their "I told you so" comments. In the first place it will be owing to their low diet and want of proper nourishment that father has got influenza, or Tommy mumps or measles—beef-fed persons *never* have these affections—(which shows what an enormous proportion of vegetarians there must be)—and in the second place, now that there is illness, you *must* fall back on beef-tea, port-wine, and other "generous diet," to get up and sustain the patient's strength. However callous or deaf you might be to the supplication for the flesh-pots from those in health, you cannot, must not shut your heart to the call of the weak or suffering.

And woe betide us if we are heretic, and the patient does not recover so quickly as we could wish (if he does, we shall be suspected of having surreptitiously called the orthodox nostrums to our aid, but that by the way), so that it behoves us to give the critical and censorious as little room for their strictures as possible.

Now, what are we to get for that erewhile *sine qua non* of the sick room,

Beef Tea?

Well, before we come to the non-flesh substitutes, which are more similar in some ways to the ordinary beef-tea, we will consider what is given in the earlier stages when the stomach rejects nearly all nourishment.

Pure Fruit Juices

can usually be retained and assimilated by the most debilitated. The refreshing and restorative properties of orange, grape, and similar fruit juices are generally appreciated, though many people hold the extraordinary belief that these are best when almost all the nourishment has been fermented out of them as in ordinary wine; but not so many even of the more advanced among us, as yet, realise the wonderful healing and anti-toxic possibilities of fresh fruits, more especially grapes. Pure grape juice has been found to act with such destructive force upon disease germs of various kinds as would appear little short of miraculous.

To prepare, press out with squeezer and strain, dilute or not with hot or cold water according to the condition of the patient. The juice of an orange to a tumbler of water makes an excellent tonic drink where there is feverishness and debility of the digestive organs, and a teaspoonful or more of lemon juice may be used in the same way.

Rhubarb Juice

is very good when made from fresh, naturally-grown rhubarb. Wipe and cut small, put in covered jar in oven or steamer till the juice flows freely. This will not be ordered where there is rheumatism or the like. For such, an alkaline beverage is wanted instead of an acid.

Celery Milk

is exceedingly good, and I claim to have discovered it for myself. Wash and trim some sticks of celery. Cut small and simmer for an hour or longer in milk and water. Bruise well to get all the goodness out, and strain through jelly-bag. When fresh celery is not to be had, celery seeds may be used. Simmer in water, strain, and add milk.

Cocoanut Milk

is also very good, and will sometimes be retained when ordinary milk is rejected. Select a juicy cocoanut, pierce a hole and drain out the milk. Break and remove from shell, and pare off the brown skin very finely, so as not to lose any of the oil. Grate or run through mincer, add two cupfuls boiling water, and beat with a wooden spoon from ten to fifteen minutes; then squeeze through a cloth or potato masher. Put the cocoanut into a saucepan with more boiling water, mash over the fire for a few minutes, and squeeze again very thoroughly. If it has been squeezed in a masher the liquor may need to be strained again through a cloth or hair sieve.

For a bland soothing drink, invaluable in practically every form of internal irritation and debility, Barley Water reigns supreme, and in its preparation Robinson's Patent Barley will be found invaluable.

Smooth one or two spoonfuls to a cream with cold water. Pour on boiling water, stirring all the while, and boil gently for five to ten minutes. When cool it will be a firm jelly, and can be diluted as required with hot or cold water, milk, fruit-juice, "Extract," &c., &c.

To come now to what more closely resembles beef-tea, we can have a liquid practically undistinguishable made from

Brown or German Lentils.

Take a teacupful of these, look over and pick very carefully so that no stones or dirt may escape notice. Scald with boiling water, and put to simmer with plenty of boiling water in a saucepan or stewing jar. Add a shallot, a bit of celery, teaspoonful ground rice, tapioca, &c., and, unless prohibited, seasoning to taste. A blade of mace, a slice or two of carrot, beetroot, &c., might be added at discretion. Simmer gently, or better still, steam for an hour. Strain, without any pressure, and serve with fingers of crisp, dry toast. Equal quantities of German lentils and brown beans may be prepared exactly as above to make Savoury Tea, as also a mixture of brown and white beans. A delicious

Invalid Broth

is made thus: — Wash well a cupful of butter peas or haricot beans and one or two tablespoonfuls pot barley. Put in saucepan or double boiler with water, and cook for two to three hours. Season and strain. Celery, onion, parsnip, &c., may be added if desired. Some milk may also be added, and, if wished specially rich and strengthening, one or two eggs beaten up. Warm up only as much as is needed at one time, and serve with toast or triscuits. Variety of flavour, &c., may be contrived by mixing lentils, dried green peas, &c., with the haricots, or instead of these, tomatoes may be sliced and added ten minutes before straining.

I need not here give recipes for ordinary oatmeal gruel, but

Lentil Gruel

may be new to some. Take a dessert-spoonful lentil flour — the "Digestive" lentil flour is always to be depended on — smooth with a little cold milk or water in a saucepan. Add three teacupfuls boiling milk or barley-water and simmer for fifteen minutes. A little extract such as "Carnos" or "Marmite" may be added to this or any of the foregoing broths.

These extracts, "Carnos" and "Marmite," are exceedingly useful in the sick-room, as they can be so easily and quickly prepared. "Carnos" being a fluid extract, is especially handy. A teaspoonful of that, or a half teaspoonful "Marmite" to a cupful boiling water makes a delightful cup of savoury tea. Be careful not to make too strong. Such extracts may also enter with advantage into

Savoury Custard.

Beat up an egg, and add to it half a teacupful milk, and either a teaspoonful "Carnos" or rather less of "Marmite," the latter dissolved in a little boiling water. Add pinch salt. Turn into a buttered cup or tiny basin, cover with buttered paper, and steam gently for seven or eight minutes till just set.

The following is a very dainty and novel

Egg Flip.

Separate the white from the yolk of an egg and beat up the white quite stiff. Beat up the yolk and add to it the strained juice of an orange or some "Nektar." Mix all lightly together and serve in a pretty glass or china dish.

White of Egg

may be made more attractive for little folk if poached by spoonfuls for a minute or two in boiling milk, and served with a little pink sugar dusted over.

Orange Egg Jelly.

Rub 2 ozs. loaf sugar on the rinds of 2 oranges till it gets as much flavour as possible, then put in a basin with the strained juice and a teaspoonful lemon juice. Bring a very small quantity of vegetable gelatine — previously soaked for an hour in cold water — to boil in a breakfastcupful of water. One-eighth of an oz. of this gelatine is enough as it is so strong. Stir till quite dissolved and strain over the sugar, &c. When cool add the yolks of two eggs beaten up, and whisk till white and frothy. Beat the whites very stiff and add them. Beat all thoroughly, and when just about to set pour into a wet mould. Or allow to set and then pile up by rocky spoonfuls in a glass dish.

When an invalid is getting past the "sloppy" stage and is able for solid nutriment

Steamed Barley

is perhaps the most valuable food of any, and dyspeptics who experience difficulty in getting any kind of food to agree would do well to go on a course of this — not for one day or two, but for weeks and months together. Wash well in cold water a teacupful of *pot* barley. Put on in clean lined saucepan with plenty of cold water, bring to boil slowly, and if there is the least suspicion of mustiness, drain and cover with clean water. When it comes a boil again, turn into a pudding basin or double boiler, cover and steam for at least six hours. Twelve hours is much better, and it is safest to put on one day, what is wanted for the next. Onions, celery, tomatoes, &c., may be added at discretion. When to be used, this barley should turn out firm enough to chew, and may be eaten with thin dry toast or "Triscuits."

Besides these home-made preparations, there are many valuable foods to be had ready for use, or requiring but little preparation, thus affording change and variety, not only to the patient, but to the nurse or cook, who must often be heartily tired of making up the same gruels and mushes for weeks or months together. The Barley Mint, Patriarch Biscuits, and Barley Malt Biscuits to be had from the Wallace Bakery, 465 Battersea Park Road, London, S.W., come in very handy. The Barley Malt Meal can be made into a gruel or porridge, while Barley Malt itself may be added to any ordinary preparation to aid digestion. Barley Malt Meal Gruel has been found a sovereign remedy for constipation, obstinate cases yielding to it when all other treatment had failed. Make in usual way and add one or two large spoonfuls treacle or honey. The biscuits may be grated and made into a mush with hot milk, &c., or they may be soaked over night in as much hot water, milk, or diluted Extract as they will absorb, and then be put in the oven to warm through. Gluten Meal is another among many valuable Invalid Foods which there is space only to mention here; while the value of Robinson's Patent Groats for gruel is widely appreciated.

For diabetic and anaemic patients there are one or two other valuable foods now on the market specially prepared to nourish and enrich the blood, while at the same time starving the disease. Barley Malt Meal is specially good, also a recent "Wallaceite" product, "Stamina Food."

The "Manhu" Diabetic Foods

are well known and highly recommended. The following

"Manhu" Diabetic Savoury

will be welcome to those whose dietary is of necessity so restricted. 1/2 pint Savoury Tea (p. 90) or diluted "Extract," 1 egg, 1 tablespoonful "Manhu" Diabetic Food, 1/2 oz. butter, salt and pepper.

Melt butter in saucepan, add the food, and mix over slow fire till butter is absorbed. Add the savoury liquid, cook for a few minutes, add seasoning, beat in yolk of egg, then the white stiffly beaten. Mix lightly. Pour into pie-dish, and bake in quick oven for 15 minutes.

* * * * *

A Realised Ideal In Food Production.

Ideal Food Reform means much more than "going without meat." It means the use of only such foods as will thoroughly nourish the body without injuring it.

For instance, most popular biscuits are made from an impoverished white flour, and raised with chemicals, which injure the system. Again, white bread is an artificial one-sided food, and is raised with yeast. Yeast is a ferment, the product of brewery vats, and is not expelled from the loaf by baking.

Thorough-going Food Reform demands bread, biscuits, &c., made with entire whole wheat flour, and free from chemicals, yeast, and other impurities. This is a high ideal: can it be realised?

It has been realised. The Wallace P.R. Foods Co. was founded expressly for-the purpose of making bread, biscuits, cakes, and other foods on scientific principles, which a great London "daily" has described as

100 Years in Advance of the Age.

In this model bakery the only flour used throughout is an entire wheatmeal ground to a marvellous fineness; and all other ingredients are the very best and purest. Chemicals, cheap fats, and yeast are banished.

Thousands have proved that the regular daily use of the P.R. Biscuits, Bread, &c., not only delights the palate, but eradicates many stubborn diseases, and brings about a steady improvement of health in cases where drugs, patent medicines, and all other unnatural methods have failed.

30 Samples of delicious Bread, Cake, Biscuits, and Coffee, 1/6 carr. paid.

Box Biscuits and Coffee only, 1/3 carr. paid.

_P.R. Specialities are stocked by all Health Food Stores.

Sole Makers:_

The Wallace P.R. Foods Co.

465 Battersea Park Rd., London, S.W.

* * * * *

INFANTILE MORTALITY

"COW & GATE" Dried Pure English Half-Cream Milk

The Superiority of Dried Milk over Fresh Cow's Milk was strikingly demonstrated by the experiments of the Sheffield Corporation Scheme for Reducing Infantile Mortality, given in a paper by ALBERT E. NAISH, M.A., M.B., B.C., Cantab., Assistant Physician, Sheffield Royal Hospital, in the September 3rd issue of the *Medical Officer*. For the purpose of these experiments our milk was used with that of two other makers.

OUR MILK BEING MADE DAILY AT OUR OWN FACTORIES

can be supplied in a much fresher condition than Foreign or Colonial makes. Besides the fact of our supplying several Infant Milk Depots and Creches, we have Thousands of Letters from grateful mothers, from all parts, who testify to the splendid results from feeding their babies on our Dried English Milk.

West Surrey Central Dairy Co. GUILDFORD.

It can be obtained of most Chemists and Health Food Stores, in Tins and

Packets, 1/1. each.

We make Dried, Full-Cream, and also Separated Milk, as well as the above.
Prices on application.

* * * * *

Savoury Gruel.

Dissolve about 1-1/2 teaspoonfuls vegetable extract—"Marmite," "Carnos," Mapleton's Nut Extract are all good—in 3 gills boiling water. Have a tablespoonful of either Gluten Meal, Barley Malt Meal, Banana Oats, &c., made smooth with a little cold water—add seasoning, a little grated onion, celery, &c.—and mix it with the "Extract" tea. Boil all together, stirring constantly for 5 or 10 minutes, then strain.

This savoury gruel may be acceptably varied from time to time by substituting Robinson's Patent Barley or Groats for the above.

Almond Cream Whey.

One pint milk, 1 dessertspoonful lemon juice, 1 tablespoonful Almond cream or Cashew nut cream. Bring milk nearly to boiling point, and add lemon juice. Let stand till it curdles. Strain and stir in the nut cream, also sweetening to taste.

"Nutter" Milk

(For Wasting Diseases, in place of Cod Liver Oil).

Put 1 oz. "Nutter," or other good vegetable fat, in small enamelled saucepan, and pour on 1/2 pint of milk. Heat very slowly nearly to boiling point. Stir or beat with wooden spoon till cool enough to drink. Pour into warm glass and sip slowly. If not all used at once, heat slowly, and mix well each time to be used.

Almond Milk Jelly.

Make up 1/2 pint almond milk by shaking up 1 tablespoonful Mapleton's concentrated almond cream with 2 gills water. Soak 1/8 oz. vegetable gelatine in cold water for an hour. Strain off the water and put in saucepan with the almond milk, rind of 1/2 lemon and juice of whole one, also 2 ozs. sugar. Stir over gentle heat till gelatine is dissolved. Strain and mould in usual way.

Onion Gruel (for a Cold).

One lb. onions, 1 apple, a little sugar, salt, ground cloves or mace, and white pepper, 1/2 gill boiling water, 2 tablespoonfuls "Cow and Gate" dried milk, 1 oz. butter or vegetable fat. Peel and chop the onions and scald with boiling water. Put on to simmer, with the apple chopped small, the water, butter, &c.—all except the dried milk. Cover and cook gently till tender. Sprinkle in the dried milk, and cook for a few minutes longer. Serve very hot.

The dried milk—full cream, half cream, or separated according to need of patient—may be added to any of the foregoing recipes where concentrated nourishment is required.

MISCELLANEOUS.

Mushroom Ketchup.

Fresh mushrooms—those just past the cooking stage for preference—spread not too thickly on flat dish. Sprinkle liberally with salt and let stand from 24 to 30 hours. Strain off liquor, pressing mushrooms thoroughly. Boil and bottle. If preferred, spices may be added, but we prefer it "unadulterated."

"Reform" Cheese.

(Mrs C. LEIGH HUNT WALLACE, London.)

The following is an original recipe for cheese without rennet given me by
Mrs Wallace, a well-known pioneer in Food Reform.

Put the strained juice of 3 lemons into a quart of boiling milk, then remove immediately and set aside to cool. Place a wet cheese-cloth in a hair sieve and place in the contents of the saucepan. Let drain, shape by gathering the cloth together, compress and leave for a little. Garnish with parsley. Eaten with raw tomatoes and oatcakes it is delicious. The whey, if sweetened to taste, forms to those who like it a pleasant, cooling, and health-giving beverage.

Manhu Wheat Yorkshire Pudding.

Three tablespoonfuls Manhu Wheat, 2 eggs, a little over half a pint of milk; salt to taste; 1 oz. butter.

Put the wheat in a basin, mix with milk until it forms a nice batter; add a little salt. Beat up the eggs very lightly, and add to the batter. Put the butter in a small baking tin in the oven, and, when hot, pour in the batter. Bake about 20 minutes in a sharp oven.

Breakfast Savoury.

Allow 1 egg, 1 small tomato, 1/4 oz. butter or vegetable butter, to each person. Scald, peel, and slice tomatoes, and fry till quite cooked in the butter. Add seasoning to taste—salt, pepper, little grated onion, pinch herbs, a little Vegetable Extract or Ketchup—any or all of these—and the eggs, which may either be dropped in or slightly beaten up. Scramble till set, and serve heaped up on hot buttered toast. A pleasing variety of flavour is produced by substituting walnut butter for the other. The toast might also be spread with a very little "Marmite."

MODEL DINNERS FOR A WEEK.

SUNDAY.
Brown Soup. Nut Omelette. Almond Custard with Stewed Fruit.

MONDAY.
Hotch-Potch. Sausage Rolls. Canary Pudding with Appel-Moes.

TUESDAY.
Clear Soup. Savoury Lentil Pie. Lemon Cream.

WEDNESDAY.
Tomato Soup. Scotch Haggis. Cobden Pudding.

THURSDAY.
Mock Hare Soup. Kedgeree. Provost Nuts Pudding.

FRIDAY.
White Soubise Soup. Sea Pie. Banana Custard.

SATURDAY.
Split Green Pea Soup. Macaroni Egg Cutlets. German Tart.

NOTE.—The above is only an outline. Vegetables, &c., will be added as they are in season.

* * * * *

FOOD REFORMERS KNOW

the difficulty experienced in starting the better way in diet. These can be overcome by dining at …

'THE ARCADIAN' Food Reform Lunch and Tea Rooms
And HEALTH FOOD STORES,

152 St Vincent St., Glasgow

(Within 2 minutes of Central Station). The most up-to-date and artistic
Food Reform Restaurant in the Kingdom.

* * * * *

ADDITIONAL RECIPES.

SOUPS.

Nut Soup.

One pint boiling water, 3 tablespoons grated walnut or walnut meat preparation, some onions sliced, spoonful gravy essence, 1/2 lb. sliced tomatoes, a little "Nutter." Make the fat hot and fry onions lightly, add sliced tomatoes and grated nuts, and stir for a few minutes. Pour boiling water over, and allow all to simmer for 20 to 30 minutes; season to taste, and serve.

Split Green Pea Soup.

One lb. split green peas, 1/2 lb. onions, 1/2 lb. carrots, 2 quarts boiling water; scald peas with hot water, and put on with the 2 quarts (8 breakfast cupfuls) boiling water, and the onions chopped small. Simmer for an hour, and add the carrot flaked or chopped small. Cook for another hour, add seasoning, herbs, parsley, &c., and it is ready for use. This is a most delicious and nourishing soup, and very quickly and easily prepared. Can be varied by using tomatoes instead of the carrots, or by the addition of any other vegetables as cauliflower, leeks, spring onions, &c., also by substituting 4 to 6 ozs. rice or barley for same quantity peas.

Simple White Soup.

One large onion, 1 large potato, 1 tablespoonful oatmeal, 1 tablespoonful butter. Boil gently 1 hour in 2 breakfast cupfuls milk and 1 of water. Pass through a fine sieve, and serve very hot. May be varied by substituting Provost Nuts or Marshall's "Cerola" for the oatmeal.

Plasmon Vegetable Soup.

Two carrots, 2 turnips, 1 leek, 1 onion, 1-1/2 oz. butter, 1 teaspoonful celery seed, 2 lumps sugar, 1 bay leaf, 1 pint Plasmon white stock, 1 oz. flour, 1 gill milk, salt and pepper. Shred vegetables into thin strips. Melt butter, and add Plasmon stock while boiling. Cook till vegetables tender. Blend flour and milk smoothly, and add gradually, also seasoning. Boil a few minutes longer. For

Plasmon Stock,

put 1 oz. Plasmon in saucepan, and add gradually half a pint lukewarm water, stirring continuously. Place over the fire, and boil for two minutes. When cold, this should be a thin, semi-transparent jelly.

Cream of Barley Soup.

Prepare a white or clear stock (p. 11), or make a hasty stock by boiling some lentils, split-peas, or haricots with a good quantity of chopped onion till of the strength required. Failing any of these, a spoonful or two of vegetable extract will do very well. Bring to boil, and season to taste. In a basin smooth some of Robinson's Patent Barley to a cream with cold water or milk, allowing one tablespoonful to the pint. Pour on to this the boiling stock, stirring all the time. Return to saucepan, boil up, and allow to simmer for at least ten minutes. More milk may be added if desired, and this soup can be varied and enriched by the addition of the yolks of one or two eggs. These should be well beaten up and put in tureen before dishing. I may say here that the Patent Barley is must useful for thickening any kind of soup, stock, or gravy.

SAVOURIES.

Nut Soufflee.

A teacup each of grated walnuts, brown bread crumbs, and milk, a beaten egg, pepper and salt. Mix well, grease a tin mould, pour in mixture, and steam for an hour. Serve with Tomato Sauce. When cold, it can be cut in slices, rolled in egg and bread crumbs, and fried a nice brown.

NOTE.—The above can be varied by using a different kind of nuts or Mapleton's Nut-meat Preparation, and by the addition of a little grated onion, minced parsley, and one or two teaspoonfuls Vegetable Extract.

Savoury Nut Omelette.

A large cup of grated walnuts or Brazil nuts, a cup of brown bread crumbs, pepper and salt to taste, a little grated onion, 2 teaspoonfuls finely chopped parsley; also 2 eggs well beaten, and a cup of milk. Mix all the ingredients together. Have ready an omelette pan with a good layer of hot fat or butter. Pour in the mixture, slowly brown on one side, cut in 4 or 6 pieces when they will be easily turned, then brown on the other side. Serve hot, with brown sauce, vegetables and potatoes in the usual way. A still simpler way is to bake in shallow baking tin in brisk oven 30 to 40 minutes. Use plenty of fat.

NOTE.—The above can be very easily prepared by using Mapleton's Nut-meat Preparation instead of the grated nuts. Walnut or brown Almond meal would be especially suitable.

Sea Pie.

Cook together a variety of tender spring vegetables—carrots, turnips, cabbage, pens, French beans, &c. First brown some onions with "Nuttene," add water with some vegetable extract—"Marmite" or "Carnos"—also some ketchup and seasoning. When boiling, add the carrots and turnips—not too small—then a fair-sized cabbage cut in four pieces, the peas shelled, or French beans cut lengthwise. The carrots and turnips should be cooking for some time before the cabbage, &c., is put in. See that there is plenty of liquid to cover, and put on the following paste:—Take four heaped tablespoonfuls self-raising flour, a piece of "Nuttene" or butter the size of a small egg. Rub in very lightly with the tips of the fingers, add pinch pepper and salt, and mix to a soft dough with a little water. Flour well and roll out lightly to not quite the size of round stewpan to leave room for swelling. Make a hole in centre, add quickly to contents of pan while fast stewing, keep lid very close, and cook for 3/4 of an hour. Serve very hot. Sea Pie may also be made with mushrooms stewed till tender, with teaspoonful "Extract" and tablespoonsful ketchup. Have plenty of liquid.

NOTE.—The above is exceedingly good, very simple to prepare, and may be varied in innumerable ways. For those who prefer to dispense with chemical raising materials, I may say that the paste is very good made with ordinary flour, or with a mixture of wholemeal and flour. An egg *may* be beaten and mixed in, but it rises very well without. The same paste can be put over any stew—German Lentil, Haricot Bean, &c.—great care being taken that there is plenty of liquid.

Scotch Oatmeal Pudding.

One lb. oatmeal, 1/4 lb. onions, 1/2 lb. vegetable suet or 1/4 lb. each of suet and pine kernels; pepper and salt. Run the pine kernels through nut-mill, and put with suet in frying-pan. When hot, add the onions finely chopped, and after these have cooked for a few minutes add the oatmeal, which should be crisp and not too fine. Cook all for some time, stirring constantly to prevent burning. Wring a pudding cloth out of boiling water, flour well, and put the oatmeal, &c., in, and tie up at each end in the form of a roll, leaving a little room to swell. Plunge in fast-boiling water, and boil for 3 to 4 hours. Turn out of cloth carefully so as not to break. It may be served as it is, but is much nicer if put in a baking tin, basted with hot fat, and baked till brown and crisp. Serve with brown sauce or nut gravy.

This may be divided into a number of small puddings. These are particularly good if allowed to cool, and then brushed over with a little white of egg before being toasted.

Hasty Oatmeal Pudding.

Make some vegetable fat very hot. Add a little onion, grated or very finely chopped, and stir till nearly cooked. Allow a teacupful oatmeal to each tablespoonful of fat, and stir in along with a little salt and pepper. Cook over very moderate heat till crisp and brown all over, turning about almost constantly as it is very ready to burn. Shredded Wheat Biscuit crumbs, Granose Flakes, or Kornules may be used in place of the oatmeal. Less fat will be required.

Walnut Mince.

Six ozs. grated nuts, 4 ozs. breadcrumbs, 1 oz. Nut butter. Make fat hot in saucepan, add nuts, and stir till lightly browned, taking great care not to burn. Add breadcrumbs and seasoning to taste—large spoonful grated onion, pinch herbs, &c.—also ketchup or vegetable extract—"Carnos" or "Marmite"—with boiling water to make up 2 gills—rather less if a dry consistency is preferred. Simmer slowly for 15 minutes. Serve with sippets of toast or fried bread. Brazil, peccan, or hazel nuts may be used instead of walnuts.

Savoury Lentil Pie.

With the help of the above mince quite a number of delicious savouries can be contrived with but little extra trouble. The following pie will be found delicious:—Wash well 8 ozs. red lentils, and put on to cook with 2 ozs. each of chopped or flaked carrot, turnip, and onion, 1 oz. butter, pinch herbs, ditto curry powder, teaspoonful sugar, and usual seasonings. Cover with just as little water as will cook the lentils without burning, and simmer or steam closely covered for about half-an-hour till lentils a thick puree. Some ketchup,

"Extract," or tomato is an improvement; add nut mince prepared as above, mix well and simmer a few minutes longer. It should be of the consistency of a thick mush. Put in pie-dish, and set aside to cool. Cover with

Batter Paste

made with 6 ozs. self-raising flour, 2 eggs, 1-1/2 gills milk, 3 ozs. butter or vegetable fat. Rub the butter into the flour, and make into stiff batter, with the eggs well beaten, and the milk. Pour over contents of pie-dish and bake till well risen and a nice brown in fairly brisk oven.

Nutton Pie.

One-and-half lbs. "Nutton," [Footnote: A very fine Nut Meat, put up by R. Winter, City Arcades, Birmingham.] cut in dice, 1/2 lb. tomatoes, 1/4 lb. cooked macaroni, 1-1/2 lbs. cooked potatoes, sliced. Dust with pepper and salt, pour in stock to within 1/2 inch of top; cover with good whole-meal crust, made with Winter's cooking "Nutbut"; bake.

Nutton Chops.

One lb. No. 1 "Nutton," minced through a food chopper, 3/4 lb. zweiback bread crumbs, 2 ozs. macaroni, cooked and finely chopped, pepper and salt to taste. Mix with egg and form into chops; use a piece of uncooked macaroni for the bone; brush with egg and bread crumbs and bake, or fry, with nutbut—this quantity should make 8 chops.

Nutton Meat for Mock Sausage Rolls.

One lb. No. 8 "Nutton," put through a food chopper, 1/2 Spanish onion boiled and finely chopped, 2 teacupsful zweiback bread crumbs, a little sage, salt to taste. Have quantity required of puff pastry, roll out and divide into squares, putting a little sausage meat in the centre, wet the edges and fold over. Place in a hot oven and bake 10 minutes to 1/4 hour.

Stewed Onions.

Select about a dozen good hard onions, as nearly of a size as possible, and weighing 6 or 8 to the lb. Make 2 ozs. or so vegetable fat—"Nutter" is very good—smoking hot in large stewpan, add the onions, and stir about till nicely browned all over; be careful not to burn; if fat not all absorbed pour it away. Cover with boiling water, add seasoning, pinch herbs, &c., cover and stew gently till cooked—about an hour. There should be a rich brown gravy, so that this makes a most appetising dish to serve with a dry savoury.

Cheese Moulds.

One pint milk, 1/2 lb. grated cheese, 3/4 lb. wheaten bread crumbs, 2 eggs, 1 teaspoonful salt, 1/4 teaspoonful mustard, 1/4 teaspoonful pepper. Put milk, cheese, and crumbs into a pan and bring them almost to the boil, add seasoning and eggs, and stir till thick, but do not let it boil. Butter some small dariole moulds and sprinkle them with some chopped parsley. Press in the mixture, dip in hot water, and turn out.

* * * * *

MAPLETON'S NUT FOODS WARDLE, LANCASHIRE.

	PER LB.
	S. D.
Walnut Butter	1 0
Cocoa Nut Butter	1 0
Cashew Butter	1 0
Almond Margarine	1 2
Nut Margarine	0 10
Blended Nut Margarine	0 10
Honey & Nut Margarine	1 0
Pea Nut Butter	0 9

Almond Cream 1 10
Hazel Cream 1 4
Cocoa Nut Cream 0 10
Nut Milk 1 4
Cooking Nutter, 1-1/2 lb. carton 0 11
Nutter Suet 0 8
Cooking Nut Oil 1 0
H.M.R. Nut Oil 1 6
Walnut Oil 2 6
Olive Oil 1 5
Salted Almonds (packet) 0 11
Blanched Almonds 1 3
Cooking Almonds 1 0
Jordan Almonds 1 8
Twin Jordan Almonds 1 2
Walnut Halves 2 0
Broken Walnuts 0 8
Pine Kernels 0 11
Roasted Pine Kernels 1 0
Pea Nuts 0 4
Roasted Pea Nuts 0 5
Blanched Pea Nuts 0 6
Cashew Nuts 0 9
Hazel Nuts 0 10
Monkey Nuts 0 4
Almond Meal 1 6
  " (Unblanched) 1 3
Hazel Meal 1 0
Walnut Meal 0 11
Chestnut Meal 0 4
Desiccated Cocoa Nut 0 5
Pea Nut Meal 0 7
Roasted Pea Nut Meal 0 7
Banana Meal 0 6
Dried Bananas 0 6
Figs 0 4
Dried Pears 0 9
Orange Peel 0 5-1/2
Lemon Peel 0 5-1/2

Citron Peel 0 9
Malted Almonds and Hazels 1 9
Cereal Cream 0 6
Nut Graino 0 3-1/2
Wholemeal (3-1/2-lb. bag) 0 6
Malt Extract 6-1/2d. and 1 0
Nut Extract 0 7-1/2
Malt Extract & Nut Oil 0 7
Powdered Dried Herbs 0 1
Gravy Essence 6d. and 1 0
Nut Gravy 1 0
Finest Honey 1 0
Finest Cocoa 2 0
Pure Coffee 1 10
Banana Coffee 1 2
Nut Coffee 1 0
Lapee Cereal Coffee 0 9
Rich Wholemeal Sultana Cake 0 10
Nut Cakes (each) 0 6
Nut Milk Chocolate 1 0
Nut Milk and Fruit Chocolate 1 0
Nut Milk Chocolate with Marzipan 1 0
Milk Chocolate 2 0
Nucolate (packet) 0 1
Honey & Nut Caramels 1 2
Toasted Corn Flakes 0 5
Dates and Nuts 0 1
Egg Beaters (each) 1 0
Nut Mill " 16 6
Nut Graters " 1 6
Unpolished Rice 2d. and 0 3

SAVOURY NUT MEATS.
S. D.
White Almond Meat 1 0
Walnut Meat 0 10
Pine Kernel Meat 0 10

Brown Almond Meat 0 10
Savoury Meat 0 10
Red Savoury Meat 0 10
White Fibrose Nut Meat 1 0
Brown Fibrose Nut Meat 1 0
Potted Tomato and Nut (tin) 1 0
Nut Meat Preparation (4 kinds)

WHOLEMEAL BISCUITS.
S. D.
Water Wheat (3 lb.) 0 11
Shortened Wheat " 1 0
Malt Wheat " 1 0
Nut Wheat 1 0
Short Wheat 0 5
Nut Wheat Crackers 0 6
Hazel 0 6
Milk 0 6
Oat Flake—Sweet 0 8
Oat Flake—Plain 0 8
Ginger Cake 0 8
Weinmost (13 kinds)
Mostelle (3 kinds)
Preserved Ginger 0 9
Hallowi Dates 0 3
Sair Dates 0 2

FRUITARIAN CAKES.
S. D.
Apricot and Nut 0 6
Pear and Walnut 0 6
Plum and Nut 0 6
Cherry and Nut 0 6
Muscatel and Almond 0 6

Almond and Raisin 0 6
Extra Rich 0 6
Cocoa Nut Sandwich 0 6
Chocolate Sandwich 0 5
Popular Variety 0 6
Raisin and Cocoa Nut 0 5
Muscatel and Cocoa Nut 0 5
Date and Orange 0 4
Date and Lemon 0 4
Date and Ginger 0 4
Date and Hazel 0 4
Date and Pine Kernels 0 4
Fig and Raisin 0 4
Fig and Citron 0 4
Fig and Ginger 0 4
Carraway 0 4
Date and Cocoa Nut 0 3
Date and Nut 0 3
Date and Walnut 0 3
Fig and Cocoa Nut 0 3
Fig and Nut 0 3
Date and Almond 0 3
Date Caramels 0 4
Fig Caramels 0 6

NUT CAKES

(In place of Cheese).

PER PKT.
S. D.

Almond 0 9
Pine Kernel 0 7
Honey and Nut 0 6
Pea Nut and Cocoa Nut 0 5

FULL PRICE LIST ON APPLICATION.

* * * * *

RODBOURN'S Health Foods Depot

40 Hanover St., Edinburgh

VEGETARIANS, or intending Vegetarians, should write or call for our List of over 400 varieties.

We have the most varied stock of Health Foods in Scotland, and can give early delivery.

Families catered for at a distance. Small orders from manufacturers are often costly. Avoid worry and save time and money by buying your goods in one lot.

NOTE.—We pay carriage up to 50 miles by goods train on 10/- orders; £2 parcels sent carriage paid anywhere.

Remember, what a wrong diet causes a right diet will cure.

RODBOURN'S, 40 Hanover Street, EDINBURGH

National Telephone. 5055

* * * * *

BREAD.

Considerable difficulty seems to be experienced in many quarters in getting really good bread free from chemicals and other deleterious matters. In some households the problem is solved by subsisting solely on certain approved kinds of biscuits, one I heard of keeping exclusively to Shredded Wheat Biscuits and Triscuits, while another stood by the "Artox" Biscuits. Besides these there are several other specially good whole-wheat biscuits, among which may be mentioned Chapman's Nut Wheat Biscuits; Winter's "Mainstay" series of Diet Biscuits, including some dozen varieties, all excellent, ranging in price from 4d. to 8d. per lb.; and the "P.R.," a Wallaceite specialty. Among the latter the "Barley Malt," "Crispits," "P.R. Wheatmeal," "New P.R. Crackers," &c., are to be specially recommended. Most people, however, prefer to have something more in the way of a loaf, and those who can make

Home-Made Bread

should have no difficulty in providing a toothsome and, at the same time, perfectly wholesome article. Directions for Wallace Egg Bread are given on page 74, and for Wheatmeal Gems, made with meal and water only, page 73. The following is a still simpler method:—Get a reliable whole-wheat flour; Hovis, Manhu, and Artox are each excellent, and will commend themselves severally to different tastes and requirements. The latter, it is useful to know, is used exclusively in the Wallace P.R. Bakery—a guarantee for its purity and wholesomeness. To prepare, take amount of flour required, and allow 1 or 2 ozs. vegetable butter or nut oil to the lb. Salt or not to taste. Rub in the butter and make into a stiff dough with cold water. Run two or three times through an ordinary mincer to aerate, and form into a long roll, but without pressure of any kind. Divide into suitable pieces or put in loaf pans, and bake in well-heated oven for 30 minutes to 1-1/2 hours, according to size. Most people will prefer small crusty loaves or rolls which get baked right through. For ordinary

Home-Made "Hovis" Bread

take 3-1/2 lbs. Hovis flour, 4-1/2 gills warm water, 1 oz. German yeast, 1 oz. salt, teaspoonful sugar. Mix salt with dry flour, dissolve yeast with sugar; make a hollow in centre of flour, put in yeast and pour on the warm water; mix well, folding in the flour from the outside to the centre, and let stand about 30 minutes in a warm place. Knead a very little, divide into small loaf pans, and allow to rise for another 15 minutes. Bake in very hot oven about 30 minutes, reduce heat, and bake 15 minutes longer. The above quantity will make five 1-lb. loaves.

CAKES AND SCONES.

The following are a few additional recipes for cakes and scones, most of which include one or other of the numerous Health Food specialties and dainties now upon the market, but which are not nearly so well known as they deserve to be.

Bruce Cake.

(Miss MACDONALD, Diplomee, Teacher of Cookery.)

1 lb. wheaten flour, 5 ozs. soft sugar, 2 ozs. butter or "Nutter," 4 ozs. sultanas, 4 ozs. currants or candied peel, 2 teaspoonfuls baking powder, 1/2 teaspoonful mixed spice. Cream sugar and butter. Add flour, fruit, spice, and baking powder. Mix with just enough water to moisten. Bake in good steady oven for about an hour.

Tweedmont Sultana Cake.

1/2 lb. butter or "Nutter," 3/4 lb. flour, 1/2 lb. soft sugar, 6 eggs, 1 lb. sultanas. Beat butter or "Nutter" to a cream, add the sugar, and beat for twenty minutes longer. Add two eggs, and beat again till thoroughly mixed, adding a little flour to prevent curdling, and repeat till all the eggs are in. Then sift in the flour, and add the sultanas cleaned and rubbed with flour. Mix lightly and pour into well greased cake tin. Bake in slow oven 1-1/2 hours.

Murlaggan Cake (Steamed).

1 cup whole-wheat meal, 1 cup flour, 1 teaspoonful ground ginger, 1 teaspoonful mixed spice, 1 cup Sultanas or stoned raisins, 2 tablespoons "Nutter," 1/2 teaspoonful baking soda, 2 tablespoonfuls syrup or treacle, or 1 of each; 1 egg, a very little sour milk. Rub "Nutter" or butter into flour, mix all dry things. Beat up egg, and add, with just enough sour or butter-milk to mix. Turn into greased pudding-bowl, and steam for about 2 hours. This should be a very light, wholesome cake, and is especially useful when one has not an oven. It may be varied to advantage, as by using Banana flour in place of the other, chopped dates or fruitarian cake in place of raisins, &c. A handy holiday cake.

Swiss Roll.

4 ozs. sifted sugar, 2 eggs, 4 ozs. Pattinson's banana cake flour, some jam, 1/2 teaspoonful Pattinson's baking powder or small teaspoonful home-made baking powder, 2 tablespoonfuls milk or orange juice. Put sugar and eggs in a basin, and switch up with "Gourmet" pudding spoon or a couple of forks for fifteen minutes. Add the milk and beat again, then the flour, previously mixed with the baking powder and sifted in. Beat all very thoroughly. Grease well a flat baking-tin, cover with greased paper, and pour in the mixture. Bake for not more than 5 minutes in very hot oven. Turn out on a paper sprinkled with sifted sugar, remove the greased paper, spread with jam or marmalade, and roll up very quickly.

Sponge Sandwich.

Prepare mixture exactly as above. Put half in well-greased sandwich tin, colour the other half pink with a few drops of carmine, and put into a second tin. Bake as before, turn out on a cloth or sieve. Spread the under side of one with either jam, marmalade, chocolate mixture, &c., and put the other one on top. Dust over with sugar, or coat with a thin icing. For this Mapleton's Cocoanut Cream is very good.

Banana Buns.

1/2 lb. Pattinson's banana flour, 1-1/2 ozs. "Nutter," 1/2 teaspoonful baking powder, 2 ozs. sugar, 1 egg, a little milk. Mix dry ingredients, rub in the "Nutter." Beat up egg, and add with a very little milk to make a rather firm dough. Divide into small pieces, flour the hands, and roll into balls. Have a teaspoonful sugar dissolved in a few drops of hot milk on a saucer. Dip in each bun, and place with sugared side uppermost on greased tin or oven plate. Bake for about 10 minutes in rather hot oven.

Banana Flour Scones.

1 lb. banana flour, 2 ozs. butter or "Nutter," 2 ozs. sugar, 1 teaspoonful baking powder, milk. Mix flour—the banana flour sold by the lb. is best—sugar, and baking powder. Rub in butter, make into a light dough with milk. Cut into small scones, and bake in good oven about 15 minutes.

These scones are exceedingly good, and quite different from those made with ordinary flour. They may be varied by adding a few Sultanas or a beaten egg.

Manhu Crisps.

1 lb. Manhu whole-wheat flour, 1 oz. cocoanut butter, pinch salt. Rub butter into flour, and make into a dough with as little water as possible; then run twice or three times through an ordinary mincer. Form into twelve or more rolls or twists with as little handling as possible, and bake in hot oven for ten to fifteen minutes.

Manhu Scones.

1 lb. Manhu Flour, 1/2 teaspoonful carb. soda (not heaped), sour milk or butter milk to make a soft dough. Bake on a girdle if possible.

Hovis Scones.

1 lb. Hovis Flour, 1 oz. nut butter, pinch salt, 1 tablespoonful treacle, 1/2 teaspoonful carb. soda, butter milk or sour milk. Mix dry things, rub in butter, add treacle and enough sour milk to make a fairly soft dough. Mix thoroughly and quickly. Roll out not too thin, and bake in good oven about 15 minutes. The treacle may be omitted.

Hovis Gingerbread.

8 ozs. Hovis Whole-Wheat Flour, 8 ozs. ordinary flour, 4 ozs. Nuttene, 8 ozs. stoned raisins, 8 ozs. treacle, 6 ozs. sugar, 1 egg, 1 teaspoonful ground ginger, 1-1/2 do. mixed spice. Melt together the sugar, butter, and treacle. Mix dry things together. Beat egg and pour hot treacle among it, then add to dry things. Mix and beat well. Pour into greased tin lined with buttered paper, and bake in very moderate oven 1-1/2 hours, or, if divided in two smaller tins, 3/4 of an hour will do. Golden syrup may be used instead of treacle, in which case use little or no sugar.

Strawberry Shortcake.

Make a good short crust (p. 75) with 1/2 lb. flour—plain, wheaten, or Banana flour, as preferred—1 oz. almond meal, and 4 ozs. "Nuttene." Roll out 1/2 inch thick, cut sharply round, flute edges, and bake in hot oven till a nice brown and crisp right through. Split open, inserting a sharp-pointed knife right round and pulling apart. When cool, cover under-half thickly with strawberries, well crushed and mixed with plenty of sifted sugar. Put on top half, dust with sugar, serve cold with cream or nut cream. Another very good shortcake is made as for "Jumbles," page 79. Add a little milk or fruit juice to mixture to make less crumbly. Bake in two sections and put strawberries between.

Scotch Oatcakes.

Scotch oatmeal, 2 ozs. nut butter to lb., pinch salt, hot water. Pat oatmeal in basin, melt fat in fairly hot water, and mix in quickly to make a stiff dough. Knead to thickness required. Bake on hot girdle, and toast in front of fire.

* * * * *

"REFORM" RESTAURANT AND TEA ROOMS,

73 North Hanover Street, EDINBURGH.

* * * * *

PUDDINGS AND SWEETS.

"Provost Nuts" Pudding.

This is one of the very best puddings I know, and will, I feel sure, be welcomed by all who wish for something at once novel, simple, and wholesome. It will be found a change both from the usual "steamed" and the familiar "milk" pudding. 4 ozs. "Provost Nuts," 4 ozs. stoned raisins, 3 ozs. sugar, 3 gills milk, 1 or 2 eggs, a little spice or flavouring. Put "Provost Nuts," raisins, and sugar in basin. Bring milk to boil, pour over, cover, and allow to stand till cool. Beat up yolks and add, also flavouring, then the whites whipped stiffly. Mix well, and bake about 45 minutes in moderate oven. This pudding is also very good steamed. Use rather less milk. The yolk and white of egg need not be separated. May be varied by substituting currants, sultanas, or chopped "Fruitarian" cake for the stoned raisins.

"Provost Nuts" Walnut Pudding.

3 ozs. "Provost" Nuts, 3 ozs. grated walnuts, 3 ozs. sugar, 2-1/2 gills (i.e., teacupfuls) milk, vanilla essence. Bring milk to boil, pour over the "Provost" Nuts, and soak till cool. Put in saucepan along with the grated walnuts, bring to boil, and simmer gently for five minutes. Remove from fire, and when cold add the beaten yolks, sugar, and vanilla; lastly the whites beaten very stiff. Mix well, pour into buttered dish, and bake for 30 to 40 minutes in moderate oven. This is by no means an expensive pudding—at least when eggs are reasonable—and is dainty enough to grace even a festive occasion.

"Hovis" Walnut Pudding

is made by substituting 4 ozs. "Hovis" Bread crumbs for the "Provost Nuts." This will not require soaking, but can be put at once in saucepan with milk and grated walnuts.

"Hovis" Fruit Pudding.

3 ozs. "Hovis" flour, 3 ozs. semolina, 2 ozs. sugar, 4 ozs. currants or stoned valencias or sultanas, or equal quantities of all three, 3 ozs. chopped nut suet or pine kernels, 2 ozs. treacle, 2 ozs. coarse marmalade (see p. 83), 1 egg, 1/2 teaspoonful carb. soda, and a little spice. Sour milk to mix. Mix all the dry things; beat egg and add, also treacle, marmalade, and enough sour milk to make fairly moist. Steam for 2-1/2 to 3 hours in basin, well greased and dusted with sugar.

Farola Pudding.

3 ozs. Farola, 4 gills milk or nut cream milk, 2 eggs, sugar, flavouring. Smooth Farola to a cream with a little of the milk. Put remainder on to boil and pour over Farola in basin, stirring the while. Return all to saucepan, and cook gently for a few minutes. Beat up eggs with sugar, remove Farola from fire, and add, also flavouring. Pour into buttered pudding-dish, and bake gently for half-an-hour, or steam in buttered mould for 1 hour.

To make Farola Blanc-Mange use only 3 gills milk, and omit the eggs.

Semolina Syrup Pudding.

3 ozs. Marshall's Semolina, 3 ozs. golden syrup, 1 pint milk. For a simple, inexpensive pudding, the following is excellent, and it will, I think, be new to many. Make the Semolina in usual way—that is, bring milk to boil and sprinkle in the Semolina as if making porridge, cook gently for a few minutes with lid on, then pour into steamer-bowl. Allow to stand till cold, then put the syrup on top, and put on to steam for about 1-1/2 hours. The syrup will find its way through, and the pudding should turn out a lovely golden brown with the syrup for a sauce. No eggs, other sweetening, or flavouring required. Farola or corn flour may be done same way.

Syrup or Treacle Tart.

Cover a flat ashet with either rough puff paste or short crust, and fill in with a mixture composed of 1/4 lb. golden syrup, 2 ozs. bread crumbs, the juice and grated rind of 1 lemon. Ornament with crisscross strips of paste, and bake in hot oven. For a homely tart make a plain paste with wheat meal, and fill in with treacle and bread crumbs.

Plasmon Custard or Blanc-Mange.

This can be made with addition of Plasmon to any of the custard recipes given, or with the Plasmon and Blanc-Mange Powders. If the latter, to each powder add 1 pint of milk. Stir till custard thickens, but do not allow to boil.

Plasmon Sweet Sauce (for Puddings).

1/2 pint Plasmon stock, 1 oz. butter, 1/2 oz. flour, 1-1/2 ozs. sugar, flavouring of lemon rind, nutmeg, cinnamon, or bitter almonds. Melt butter; remove from fire, and mix in flour till smooth. Add

Plasmon stock gradually, cook for a few minutes very gently, then add flavouring. Very good with stewed fruit or any steamed pudding.

HEALTH FOOD SPECIALTIES.

This is an age of seeking after health, and many and various are the means proffered to that end. Drugs, serums, medical and surgical appliances, baths, waters, fearfully and wonderfully conceived methods of exercise, rigid and drastic schemes of dieting, &c., &c., crowd upon each other's heels until the prevailing idea in the mind of any one seeking to solve the health problem is one of hopeless mystification. Life would be too short to give them all a fair trial, even if any one could be found either foolish or courageous enough to attempt the task (I believe some *do* try everything by turns but nothing long), so one is driven perforce to make a selection; and while dismissing nine-tenths of the nostrums urged upon us as unworthy of any sane and rational consideration, we know the truth lies somewhere, and will be found by those who seek it on simple, common-sense lines. Doctors differ like the rest of us, but there is a broad general ground of agreement upon which we can all go, namely, that cleanliness, in its widest sense, including pure air, food, and water; plain, easily-digested, nourishing food; with rest and exercise in proper proportion, are the main essentials for right living, and so furnish the key to the problem. No one of these is of itself sufficient. All are necessary and inter-dependent, and it is the want of recognising this principle which so often leads to failure and consequent abandonment, or even wholesale denunciation, of the regimen followed. Thus a person may be advised to adopt certain foods, the rules and regulations regarding which he follows to the letter, but acts unhygienically in other ways, as by shutting out the fresh air, inattention to cleanliness, over-exertion or want of sufficient exercise, eating when exhausted, and so on. The food, at least if it has gone in any way against the inclination or prejudice, will of course be blamed, while really it may be quite innocent.

One might multiply instances to show how so many not only fail to find health by their unreasonable methods, but bring ridicule and disrepute on certain of the measures followed. There is no need to waste further time, however, in demonstrating the obvious. One would hope that all readers are genuinely interested in health principles, and sufficiently in earnest to promote these intelligently.

Our business in these pages lies with the food question, and in this chapter I purpose to deal specially with

Health Foods,

of which there are a large and ever-increasing number now upon the market. How people can complain of want of variety with such a seemingly endless category to choose from passes my comprehension, for the difficulty I find is to do justice to even a small proportion of them. If one were to sample a different dish every day it would take months to get over them, and great as is the outcry in these days for variety, I do not think this constant chopping and changing by any means desirable. As I have been at some pains to find out a number of really reliable Health Foods, and can speak of these from personal experience, the information given in this chapter may serve as a guide to their selection, and save considerable time and trouble. I may say that I am indebted to a number of friends and others with whom I am in correspondence for the benefit of their experience, as well as my own. It is always good to have as wide a consensus of opinion as possible, for one finds that tastes and ideas regarding the merit of the several articles vary with the individual, and with the conditions under which used.

It is difficult to know where to begin when so much claims attention. Perhaps the class of foods which have come most largely into the public eye of late years are the so-called

Breakfast Foods,

consisting generally of cereals, pro-digested or so treated as to be easy of digestion. Several of these, such as Shredded Wheat Biscuits, have been frequently referred to in different parts of the book, so that no further words are needed to commend them. If any are sceptical, or even curious, regarding "what they are," a demonstration recently described by a Manchester friend might serve to reassure them. It was quite on the American "pig and sausage" lines, for one saw the whole wheat grain going in at one part of a machine and coming out at another in the form of a "Triscuit" ready for use.

Among other specially good foods are

Granose Flakes.

These consist of the entire wheat-kernel in the form of delicious, crisp flakes, ready for use, with cream, stewed fruit, &c., or in any way in which bread crumbs may be used. They are very handy to have in the general storeroom to sprinkle over cauliflower or any dish served *au gratin*. That they are at once nutritious and easily digested is attested by the fact that physicians of high standing put their patients on a diet of "Granose." I have known personally of cases of extreme gastric debility where the patients were put on this food almost exclusively for months together.

They may also be had in the form of

Granose Biscuits,

and these are excellent for general use. Toasted for a few minutes and then buttered—or the butter may be put on while toasting—they furnish a delicacy which few will fail to appreciate.

Avenola, Toasted Wheat Flakes, Nut Rolls, and Gluten Meal, containing 30 per cent. to 60 per cent. Gluten, are among the other products of the same firm—the International Health Association, Stanborough Park, Watford, Herts—which I have space here only to name.

In the chapter on Breakfast Foods and elsewhere the various products of the
London Nut Food Co., 465 Battersea Park Road, London, S.W.—Grain
Granules, Gluten Meal, &c., are mentioned, besides which they have a
great variety of

Nut Cream Rolls and Nut Cream Biscuits,

made from pure wheat meal and shortened with nut butter. They are aerated and free from yeast and chemicals. In the way of

PORRIDGES,

I should like to specially commend

Banana Oats

as being something quite new and appetising. It is very easily prepared, requiring only about 10 minutes' cooking. It is put up in threepenny packets, with which full directions for cooking are given. I may say that I generally make of a stiffer consistency than quantities given, and cook longer in double boiler.

Another good porridge for those who cannot take the regular oatmeal can be made with

Robinson's Patent Groats.

This is best, to my thinking, when made as under:—Smooth two or three tablespoonfuls groats in a basin with a little milk or water. Pour on boiling milk or water—a cupful to each spoonful of groats—stirring the while. Return to saucepan and cook gently for 10 to 15 minutes, or in double boiler for about half an hour.

Manhu Wheat or Barley Porridge.

Take 1 part of the flaked wheat or barley to 2 parts water. Have the water boiling and salted to taste. Add the cereal all at once, and boil for 5 minutes; only stir sufficiently to keep it from burning. It may now be served, but is better if steamed half an hour or so longer in double boiler. Serve with milk or cream and sugar, or salt as preferred. When served with stewed fruit this makes a very wholesome dish. A mixture of the wheat and barley makes a very good porridge.

The value of

Provost Oats

for porridge is too well known to need comment here. I would only remind everyone that Provost Oats are prepared from the finest Scotch grain, and Scotch oats are the finest in the world. But Provost Oats is not the only product upon which Messrs Robinson & Sons rest their fame. More recently they have put upon the market a very fine cereal food known as

Provost Nuts.

This is a highly concentrated and nutritious and sustaining food, but can be digested very easily, and so is suitable in one form or other for every one. It is a grain food scientifically prepared from a combination of wheat, barley, and malt. Being cooked and ready for use it may be served simply with a little cream, milk, or stewed fruit; or cyclists or other travellers may munch them dry, and so compass the simple life right away. Besides *au naturel*, however, they may enter with advantage into quite a variety of dishes—to thicken and enrich soups, to take the place of bread crumbs in savouries, and to contrive quite a number of new and excellent puddings. Recipes for the latter are given, p. 108, and I am sure they need only be tried to become first favourites.

Kornules

are a somewhat similar preparation, and can be used in the same way.

* * * * *

HEALTH FOODS DEPOT and REFORM FOOD RESTAURANT.

RICHARDS & CO., 73 N. Hanover St., EDINBURGH.

* * * * *

NUT BUTTERS.

It will soon be impossible to even enumerate the many excellent varieties of Nut Butters and vegetarian fats upon the market. One of the first really good fats available, and one which has stood the test of time and competition, is

Cocoa Nut Butter,

put up by the London Nut Food Co., one of the earliest and most enterprising firms to whom we are indebted for doing so much to make easy the path of food reform. This is a hard white fat, very pure and sweet, suitable for use in place of cooking butter, lard, or dripping. It is especially good for frying all kinds of cutlets, fritters, &c., and being of a firm consistency, can be flaked in a nut mill or grater to be used in place of suet. In baking also it will be found very convenient to flake in this way, as it only requires to be stirred through the flour, instead of the more tedious process of "rubbing in." To

Mapleton, Manchester,

belongs, I think, the credit of producing the first really dainty and palatable

Table Nut Butters,

and his enterprise, we are glad to see, is justified by his success, he having recently acquired land, works, plant, &c., in the country, where the manufacture of the various nut foods can be carried on under ideal conditions. This must appeal to all food reformers, who realise that clean, dainty food cannot be produced amid dirty, insanitary surroundings.

Mapleton's Table Nut Margarine

(as these goods which resemble butter, and yet are not dairy butter, must now be called) is of remarkable purity and excellence, a north country dairy farmer declaring that he would not have known it from good fresh butter! Readers will sympathise with the manufacturers of pure foods who are, in obedience to an arbitrary Act of Parliament, obliged to label their goods "Margarine." It is a comfort, however, to know that the name is all these goods have in common with the often objectionable fats which come under this comprehensive title.

The Nut Cream Butters

are for table use also. They have the distinct flavour of the nuts from which prepared—walnut, almond, hazel, cocoanut, &c. The latter is, I believe, an exclusive specialty, and is useful in practically every variety of cakes, scones, puddings, and sweets. It supplies the place both of butter and flavourings. Recipes for Cocoanut Sauce, Cocoanut Icing, Cocoanut Custard, &c., will be found in the book, but it can be used in any other recipes at discretion.

Cooking Nutter, a soft, white fat, and Nutter Suet, a hard make suitable for baking, are among the other notable products of this firm.

Nuttene,

manufactured by Messrs Chapman, Liverpool, is another fat of undoubted excellence. It can be used in all departments of cookery in place of lard, dripping, suet, or butter. This firm also produces Cashew, Walnut, Almond, and Nut Table Butter of great delicacy and fine flavour.

Especially worthy of mention are the various Nut Butters manufactured by

R. Winter, Birmingham.

They are put up in several varieties—Nutarian Almond Margarine, Nutarian Walnut Margarine, Nutarian Cashew Margarine,

Nutarian Table Margarine, Nutarian Cocoanut Margarine, and Nutarian Lard for cooking. There are no finer butters on the market, and as this firm sends a 5s. parcel of their goods carriage paid one can easily sample them. These Nutarian Butters are put up in 1/2 lb. and 1 lb. carton tins—an exceedingly handy form. Cashew Nut Butter, 6-1/2d. per 1/2 lb., 1s. per 1 lb., is a first favourite.

Quite a different class of Butters, but equally valuable in extending the resources of food reformers, are those put up by the International Health Association.

Almond Butter

is very suitable for invalids and those of weak digestion. It is light, delicate, and nourishing, and can be diluted to use as a butter, cream or milk. The

Nut Butter

is made from cooked nuts only, and may be added to soups and savouries of every description with advantage both to nutrition and flavour. It contains all the valuable properties of the nut—proteid as well as fat.

Mapleton's Brown Almond Butter is also very useful in enriching soups, gravies, &c.

* * * * *

For Goods of Guaranteed Purity send to

Richard & Co.'s Health Food Stores,

73 North Hanover St., EDINBURGH.

* * * * *

NUT MEATS.

Perhaps the greatest development of all in the way of extending the vegetarian bill of fare has been in the manufacture of nut meats. Every year sees a number of new and improved preparations put upon the market, so that there is now a very large variety to choose from. All these meats can be made use of in many ways-sliced and fried, in stews, curries, &c.

The London. Nut Food Company's are well known and of undoubted excellence. There are several kinds—Meatose, Vejola, Nutvego, &c.—all quite distinctive in flavour and suited to different tastes. Certain of these contain pea nuts, the flavour of which is objectionable to some, while others give such the preference. The

F.R. Nut Meat,

however, is free from pea nuts, and is a general favourite. It is now made up with pine-kernels, and when I served it up lately, one of those partaking of it with great relish would scarcely credit its being other than a galantine of veal. [Recipes—page 40.]

Protose, Nuttose, Nuttolene, &c.,

put up by the International Health Association, Birmingham, are of a high standard of excellence. Protose will appeal to those who like the ordinary "meaty" flavours, for it is practically undistinguishable from meat. It is very good in pies, fritters, &c. The following is a favourite recipe.

Protose and Macaroni Pie.

Blanch 3 ozs. macaroni in salted boiling water for 20 minutes. Put half in bottom of buttered pie-dish and add a little seasoning—pepper, salt, grated onion, &c. Put on a layer of Protose cut in small pieces, and repeat with macaroni, seasoning, and Protose. Fill nearly up with gravy or diluted "Extract," and cover with rough puff paste (page 75).

Quite a different type of "meats" are those put up by Chapman, Health Food Stores, Liverpool. They are exceedingly tasty and appetising, and being free from any peculiar flavour, will appeal to the popular taste for "Savoury Meats." There are some 5 or 6 varieties, among which I would specially recommend "Lentose"—a vegetable brawn. Walnut meat is also very fine. They are fully seasoned, and may be used hot or cold, and are excellent when sliced and lightly fried and served with fried tomatoes, tomato sauce (page 68), or brown gravy (page 68). Another point in favour of Chapman's "Meats" is that they are put up in air tight glass moulds.

Messrs Mapleton, Manchester, also prepare several Nut and other meats, quite different, again, from any of the foregoing. They also are mostly put up in glass moulds. But the production of this firm to which I would call special attention is the

Nut Meat Preparations,

whereby one can with very little trouble contrive Nut meats for one's self. There are four different kinds—walnut, white, and brown almond (free from pea nuts), and another containing pea nuts. This preparation is in the form of a meal, and consists of grated nuts blended with certain cereals, &c. These preparations can be used in place of grated nuts in all the dishes where these form an item. (See pages 38, 39, 99, &c.)

"Pitman" Savoury Nut Meat

bears a name which guarantees its excellence. It is free from pea nuts, and is put up in 1/2-lb., 1-lb., and 1-1/2-lb. tins.

Quite the biggest development of the last year or two in this direction are the nut meats manufactured by

R. Winter, Birmingham

of "Pure Fruit Food" fame. They are put up in no fewer than nine varieties—all excellent—but of distinctive flavours. Nos. 1, 2, 3, 8 and 9 are known as

Nutton.

These are very savoury, do not contain pea nuts, are very rich in proteid, and therefore exceedingly nourishing. They comprise

Blended Nuts, Almond, Cashew, Pine Kernel, and Walnut. Nos. 4, 5, and 6 are classed as

Legumon.

These are very fine pea nut meats, and are of three different kinds—"savoury," "plain," and "fibrine." All of the above are put up in sample tins (3 1/2d.), 1/2-lb., 1-lb., 1-1/2-lb., and 4-lb. tins. A range of sample or 1/2-lb. tins (the latter cost from 5-1/2d. to 7d.) could be had for but little outlay, and would make a very welcome addition to the store cupboard. Several very good "Nutton" recipes are given (p. 102), and other ways of utilising these "meats" will suggest themselves to the practical housekeeper. They are also very good cold with salad or vegetables, and so form a handy stand-by in hot weather.

FRUITARIAN CAKES.

These are another luxury which has been added to the Reform bill of fare within the last year or two, but they are one which will appeal equally to the "unregenerate." Of these, also, there is a practically unlimited variety, and it would seem as if every month or so added some novelty to the number.

It is not possible even to name the different kinds, but they are mostly alike in being composed of uncooked fruits and nuts, thoroughly cleaned and free from stones, skins, &c., but otherwise in their natural state. They are compressed into small cakes or slabs, and put up in a handy size for the pocket—about 1/2-lb.—and also in small penny cakes.

The "Pitman" Co. Birmingham—the largest health food dealers in the world, by the way—have no fewer than 20 varieties of these cakes, some put up in wafer form. They also supply 12 samples post free for 8d., and those who are as yet unacquainted with these dainties should lose no time in sampling them. For a cyclist's luncheon there could, be nothing more suitable than the "Bananut" outfit put up by this firm, consisting of these fruitarian cakes, chocolate, banana biscuits, &c., and all for the modest price of 6d.

The London Nut Food Co.

have several varieties of very dainty small fruit and nut cakes covered with chocolate, especially suitable for a dessert sweet. Very nice also for a "pocket" luncheon.

Mapleton, Manchester,

has no fewer than 25 varieties of fruitarian cakes, put up in 1/2-lb. packets ranged from 3d. to 7d. each, also in penny packets. The "Pear and Walnut," "Apricot," &c., are very fine. Those put up by

Chapman, Liverpool,

are somewhat different from the others, but especially good. They are of different varieties of fruits and nuts, and iced over with chocolate, &c., and some as Italian Pine stuck over with pine kernels. The "Swiss Milk" Cake, a new one, is as toothsome as it is nutritious and sustaining.

* * * * *

VISITORS TO EDINBURGH SHOULD PATRONISE The New "REFORM" LUNCHEON and
TEA ROOMS,

73 NORTH HANOVER STREET.

* * * * *

BEVERAGES.

Those who find ordinary coffee too stimulating, or otherwise unsuitable, may be glad to know of some of the good cereal coffees now to be had. They strongly resemble coffee in appearance and flavour, are very refreshing and appetising, but are free from caffeine, and quite innocuous. They are prepared by a certain roasting and grinding process from various grains, so that their source is both simple and wholesome. Caramel Cereal, prepared by the International Health Association, is one of the best, as I believe it is one of the oldest, on the market. Sip It (London Nut Food Co.) is also excellent; while yet another is Lapee, prepared by Mapleton, Manchester. These, while similar in nature and composition, differ somewhat in flavour, so that various tastes can be suited. They can be prepared as ordinary coffee, but are, I think, better to have a few minutes' boiling. Full directions are, however, given with each. Mapleton has recently added Banana Coffee and Nut Coffee—both very good.

Fruit Syrups, Wine Essences, &c.,

belong to a different order of beverages. Those of Messrs Pattinson are of undoubted excellence. Their Botanic Beer, Ginger Beer Essence, Fruit Syrups—Raspberry, Black Currant, &c.—are all specially good. They are, besides, most useful in the store cupboard. Diluted at discretion, they may be used in the composition of trifles, mince-meat, puddings, &c., in place of the Sherry or other wines which are now nearly as out of date as they deserve to be, and will certainly find no place in the menage of the "Reform" housekeeper.

Another valuable accession to "Reform" Beverages has come in the shape of

Vegetarian Extracts.

These closely resemble meat extracts in appearance and taste, but are much finer and more delicate in flavour. Their source—from nuts or grains—also ensures such purity and wholesomeness, both for the article itself and for everything and everybody concerned in its manufacture, as is impossible with animal products.

"Marmite" and Carnos have been so often quoted in recipes as to need no further mention. "Vigar" Extract (Pitman Co.) and Nut Extract (Mapleton) are others among the noteworthy substitutes for Meat Extracts.

MISCELLANEOUS

There are several excellent Health Foods yet to be mentioned, but which do not come easily within any table of classification. Among the many elixirs for health-made-easy, which medical and scientific research have lent their aid to obtain, is that of a pure albumen in easily assimilable form.

Plasmon

has a world-wide reputation, and is extensively used both in medical treatment and in the domestic menage wherever it is desirable to administer nourishment without taxing the digestive organs. It is especially valuable in cases of gastric catarrh or ulceration. Recipes for Plasmon Jelly, &c., will be found pp. 98, 110, &c.

Though in the near future dairy products may be largely superseded by those of the nut family, there are still many who will prefer ordinary cow's milk, if only that can be obtained pure, free from germs, and unadulterated. Such is to be found, we are glad to learn, in the Sterile Dry Milk supplied by the

West Surrey Dairy Co.,

who have succeeded, after much careful experiment and testing, in producing milk which in the process of preparation has been deprived of no element save germs and water. The simple addition of warm water, therefore, is all that is needed to restore it to the condition of new milk. Having lost nothing of its nutritive value, grape sugar, or organic salts, it forms a safe and valuable food for infants, and should do much to lessen the dangers of feeding by hand. It may be had Full cream, Half-cream, or "Separated," so that the most delicate digestion can be suited. Besides its use for infants and invalids, it can enter into the composition of any food where milk is ordinarily used, or where additional nutriment is desired. It may be added either dry or diluted—as most convenient. One strong point in its favour is that there is no danger of its turning sour or going bad in any way—the constant danger with fresh milk; but, of course, only the quantity required for immediate use should be diluted at one time. This Milk Powder, also compressed Tablets, can be got from all Health Food Stores, as also from most grocers and warehousemen. If any difficulty, it can be had from Headquarters, in small packets at a trifling extra cost, and in larger quantities carriage paid.

"Wallacite Reg. 'P.R.' Specialties."

In various parts of book, readers will have noticed commendatory reference to several "Wallacite" goods, and I would here urge that all seeking a pure, wholesome dietary in health or sickness, should give them a trial. The range of foods is practically unlimited, every requirement of health or palate being suited, but all alike composed of pure, wholesome ingredients, guaranteed free from such deleterious substances or adulterants as yeast, chemicals, artificial colouring matter, mineral salt, &c. The variety of biscuits and cakes ranges from the plainest sorts, to suit the dyspeptic or ascetic, to the most delectable dainties for afternoon tea, not forgetting Oaten Short-cakes to specially delight the "Canny Scot." Nor need any one be at a loss to obtain supplies, for, besides the various Health Food Depots mentioned (see inside front cover), customers can obtain 5s. worth of cakes and biscuits carriage paid to any part of the United Kingdom, direct from headquarters, 466 Battersea Park Road, London.

Besides the "Bakery" products there are many additions to one's resources generally. There is "Stamina" Food for infants invalids, and, curiously enough, athletes. It is exceedingly palatable for general use in puddings, pancakes, &e., while gruel can be prepared in a few minutes. Use one part "Stamina" Meal to four parts of fast-boiling liquid, stock, milk and water, &c.; simmer five minutes, and it is ready.

In the Pale Roasted Coffee one has coffee at its best, without the harmful properties of the ordinary article. Thus, with a selection from the other "P.R." dainties, including some pure fruit preserves, cocoanut or raisin nut cheese, &c., &c., one can have not only a "Physical Regeneration Breakfast Table," but a "P.R." store-room complete in itself.

There are many other Health Foods, &c., to which one would like to call attention, but space admits of only one—Nut Oil with Extract of Malt ought entirely to supersede the cod liver oil horror. Since a much larger percentage of nut oil can be incorporated—30 per cent. or over, as against 10 per cent. to 15 per cent., which is the most that can be tolerated of the nauseous cod liver oil—its tonic and up-building properties are much greater. Any chemist will compound it, but it can now be had ready for use from Messrs Mapleton at the very low price of 7d. per lb. See price list, p. 103.

With regard to obtaining regular supplies of Health Food Specialties, no one need be at any loss. A post card to any of the leading depots will bring a price list from which to order direct. Some firms—Chapman, Liverpool; Winter, Birmingham: "Pitman" Stores, &c.—send quite small parcels—5/-upwards, carriage paid.

The "Pitman" Reform Food Stores, Birmingham, stand unrivalled for extent and completeness. Besides their "Vigar" specialties and every possible variety of Health Foods, they have an unlimited range of cooking utensils, nut mills and appliances of every kind to facilitate the wholesome preparation of food. The "Pitman" Steam Cooker is a marvel of cheapness and excellence, consisting of deep boiler and three upper compartments, whereby four different dishes can be cooked to perfection, each retaining its full flavour and nutritiveness.

One is here reminded that there are other factors essential to right, sound, healthy living besides good well-cooked food. It is desirable to have cleanliness and purity all round; and we are glad to be independent, even in the matter of soap, of the filthy refuse fats so often used in its manufacture. In this connection the following tribute to a vegetarian soap appeals to readers.

* * * * *

From "PAPERS ON HEALTH" by Prof. KIRK, of Edinburgh.

This book should be in every home; an invaluable book of reference. From all Booksellers, 3/-.

Chapped Hands.—Our idea is that this is caused by the soda in the soap used. At anyrate, we have never known anyone to suffer from chapped hands who used M'Clinton's[*] soap only.

It is made from the ash of plants, which gives it a mildness not approached by even the most expensive soaps obtainable.

If the hands have become chapped, fill a pair of old loose kid gloves with well-wrought Lather (*see*), putting these on just when getting into bed, and wearing till morning. Doing this for two or three nights will cure chapped, or even the more painful "hacked" hands, where the outer skin has got hard and cracked down to the tender inner layer.

Bathing.—Cold Baths, while greatly to be recommended to those who are strong, should not be taken by anyone who does not feel invigorated by them. As everyone should, if possible, bathe daily, the following method is worth knowing, as it combines all the advantages of hot and cold bathing. The principle is the same as explained in "Cooling" in heating. Sponge all over with hot water and wash with M'Clinton's[*] soap; then sponge all over with cold water. No chilliness will then be felt. Very weak persons may use tepid instead of cold water. These baths taken every morning will greatly prevent the person catching cold.

Cold bathing in water which is hard is a mistake, especially in bathing of infante. The skin under its influence becomes hard and dry. Warm bathing and M'Clinton's[*] soap will remedy this.

Eczema.—Skin eruptions known under this name have very various causes. Treatment must vary accordingly.

Where the cause is a failure of the skin to act properly, the whole skin of the body, especially the chest and back, will be dry and bard. In this case apply soap blankets.

If the soap blankets be too severe on the patient, then apply general lathering with M'Clinton's[*] soap. Use a badger's-hair shaving brush, and have the lather like whipped cream, with no free water along with it. We have known a few of these applications cure a case of long standing.

Where general debility is present along with the disease, use all means to increase the patient's vitality. Simple diet is best, and abundance of fresh air within and without the house by night and by day.

[Footnote *: *If not stocked by the local grocer, samples of toilet, shaving, and tooth soap can be had from the Makers, M'Clinton's, Donaghmore, Tyrone, Ireland, on receipt of 3d. to cover postage, or a large assorted box will be sent post free for 2/6.*]

* * * * *

Winter's Health Foods and Specialities

NUTTON. — The Best Nut Meat, made in six varieties, and can be used in every way in which butcher's meat is used. Recipes with each tin. 7d., 1/-, 1/5 and 3/8 per tin.

NUTTON A LA IDEAL HOME. — These delicious dainties were served recently at our stand at the Ideal Home Exhibition, Olympia, London. (See as under, page 124.)

NUXO. — A delicious savoury preparation of Nuts for Gravies and Sauces, and also makes rich and nourishing Soups. 3d. and 1/- tins.

NUTARIAN LARD. — A pure Vegetable Fat for cooking purposes; formerly known as Cooking Butnut. 1-1/2-lb. cartons, 11d.; 3-lb. cartons, 1/9; 28-lb. boxes, 10/-

WINTOX. — A pure Vegetable Product, intended to take the place of all
Meat Extracts and Beef Tea preparations. In bottles, 1/6 each.

PRUNUS. — The rapid flesh-former — self-digestive, delicious, 86% nutriment. In tins, 3d. and 1/3 each.

PRUNUS PERFECT FOOD. — The same as above in dry powder form, 96% nutriment. In tins, 3d. and 1/- each.

NUTROGEN. — A valuable Nut and Milk Food — self-digestive. In tins, 3d. and 1/- each.

NUTARIAN CAKES.

NUTARIAN MARGARINE (formerly known as Nut Butters), made in five varieties.

Mainstay Biscuits, Malt Oat Cakes, Malted Barley Cakes, Fruit Caramels,

Nutchoo, Nutarian Chocolates and many other lines.

Send for Price List and name of nearest Agent to Sole Manufacturers:
R. WINTER, Limited, Pure Food Factory, BIRMINGHAM

Nutton a la Ideal Home.

INGREDIENTS—1 lb. Nutton (No. 1 or No. 8), 1 tablespoonful flour, 1 small onion, Nutarian lard, seasoning, 1 teaspoonful Wintox.

MODE—Chop onion and fry in small saucepan; make into thick gravy with flour
and Wintox; add to the Nutton, previously chopped; form into small cutlets.
Brush with beaten egg, dip in bread-crumb, and cook in a pan of boiling
Nutarian lard.

* * * * *

CONTENTS.

SOUPS—
  Almond Milk
  Asparagus
  Brown
  Brown Sonbise
  Brazil
  Butter Peas
  Chestnut
  Cauliflower
  Celery
  Clear Soup a la Royale
  "Digestive" Pea
  Green Pea
  German Lentil
  Haricot
  Hotch-Potch
  Julienne
  Mulligatawny
  Mock Cock-a-Leekie
  Mock Hare
  Parsnip
  Palestine
  Pea ("Reform")
  Stock
  Spring Vegetable
  Scotch Broth
  Turnip
  Tomato
  Velvet
  White Soup
  White Windsor
  White Sonbise

Westmoreland

SAVOURIES—
  Artichoke Fritters
  Asparagus Cream
  Asparagus Quenelle
  Cauliflower Fritters
  Celery Fritters
  Celery Egg Cutlets
  Celery Souffle
  Celery Cream
  Dahl
  Dresden Patties
  Esau's Pottage
  Fifeshire Bridies
  German Lentil Soup
  German Pie
  Golden Marbles
  Haricot Pie
  Haricot Ragout
  Haricot Kromeskies
  Haricot Croquettes
  Irish Stew
  Kedgeree
  Leeks (Stewed)
  Mock Sole
  Macaroni Omelet
  Macaroni Cutlets
  Macaroni Mould
  Macaroni Timbale
  Mushroom and Tomato Pie
  Mushroom Patties
  Poor Man's Pie
  Rice and Lentil Mould
  Roman Pie
  Rice (Casserole)
  Rissoles
  Rolled Oats
  Savoury Brick

Sausages, Sausage Rolls
Scotch Haggis
Scotch Stew
Tomato and Rice Pie
Toad-in-a-Hole
Vegetable Goose
Vegetable Roast Duck
Vol-au-Vent

NUT SAVOURIES—
Brazil Omelet
Brazil Souffle
Brazil Quenelles
Curried Nut-Meat
Mock Chicken Cutlets
Walnut Pie

CHEESE SAVOURIES

BREAKFAST DISHES—
Bread Fritters
Bread Cutlets
Cheese Fritters
Craigie Toast
Grain Granules
Mushroom Cutlets
Nutgraino
Omelets
Pancakes (Savoury)
Porridge
Shredded Wheat Biscuits
Triscuits
Tomatoes (Stuffed)
Wheatose

EGG DISHES

COLD SAVOURIES—
  Brawn
  Legumes en Aspic
  Mock Calf's Foot Jelly
  "Reform" Mould
  Raised Haricot Pie
  Tomato and Egg
  Vegetable Mould

POTTED SAVOURIES

SANDWICHES

VEGETABLES

SALADS

SAUCES—
  Apple, Almond
  Bread
  Brown
  Caper
  Celery
  Cocoanut
  Curry
  Custard Whip
  Dutch
  Egg, Horse Radish
  Lemon
  Mayonnaise, Mint
  Mustard
  Onion
  Parsley
  Piquante
  Sweet-White
  Tomato
  Tarragon

White
Walnut

BREAD—
  Aerated, Home-made, "Hovis"
  Wheatmeal Gems
  Wallace Egg Bread

PASTRY

CAKES AND SCONES—
  Afternoon Tea Scones
  "Artox" Seed Cake, Shortbread
  "Artox" Gingerbread
  "Artox" Scones, "Artox" Tea Biscuits
  Cocoanut Cream Scones
  Dinner Rolls
  French Layer Cake
  German Biscuits
  Gingerbread, Jumbles
  Orange Rock Cakes

PUDDINGS AND SWEETS—
  Almond Custard
  Sponge Cake
  "Artox" Queen Pudding, Appel-Moes
  Banana Custard
  Canary Pudding
  Cobden Pudding
  Cocoanut Cream Custard
  Lemon Cream, Lemon Sponge

JAMS AND JELLIES

BEVERAGES

INVALID DIETARY

MISCELLANEOUS—
  Batter Savoury
  Breakfast Savoury
  Glaze
  Icing
  "Manhu" Porridge
  "Manhu" Yorkshire Pudding
  Mushroom Ketchup
  "Reform" Cheese
  Tomato Aspic

ADDITIONAL RECIPES.

SOUPS—
  Cream of Barley
  Nut Soup
  Plasmon Stock and Vegetable Soup
  Simple White, Split Green Pea

SAVOURIES—
  Cheese Moulds
  Hasty Oatmeal Pudding
  Lentil Pie with Batter Paste
  Mushroom Pie
  Nut Souffle, Nut Omelette
  "Nutton" Pie, "Nutton" Chops
  "Nutton" Sausage Rolls
  "Nutton" a la Ideal House
  Oatmeal Pudding
  Protose and Macaroni Pie
  Sea Pie
  Shepherd's Pie
  Stewed Onions

Walnut Mince

CAKES AND SCONES—
 Banana Buns, Scones
 Bruce Cake
 "Hovis" Scones, Gingerbread
 "Manhu" Crisps, Scones
 Murlaggan Steamed Cake
 Oatcakes
 Sponge Sandwich
 Strawberry Shortcake
 Sultana Cake
 Swiss Roll

PUDDINGS AND SWEETS—
 Farola
 "Hovis" Fruit, "Hovis" Walnut
 "Provost Nuts" Pudding
 "Provost Nuts" Walnut Pudding
 Plasmon Custard and Sauce
 Semolina Syrup Pudding
 Syrup or Treacle Tart